MÉDECINS SANS FRONTIÈRES

Médecins Sans Frontières (MSF)/Doctors Without Borders is the world's largest independent organisation for emergency medical relief. In its first 25 years of action, MSF has established a world-wide reputation for providing quick and effective aid to those whose lives are put at risk by war or natural disaster.

MSF was founded in 1971 by doctors determined to offer emergency assistance wherever wars and other disasters occur in the world. Its guiding principles are laid down in a charter to which all members of the organisation subscribe.

MSF is completely independent of all governments and institutions, as well as all political, economic and religious influences. Because of this policy, MSF can provide medical aid directly and effectively where other agencies sometimes cannot, with or without government consent. To ensure this independence, MSF raises at least half of its annual budget from private donors.

Every year, around 2,000 MSF volunteers work alongside several thousand local staff in projects in some 70 countries world-wide. In addition to providing emergency relief aid, they are prepared to speak out about systematic abuses of humanitarian law and human rights they witness, often bringing injustice and malpractice to world attention.

MSF's international network is made up of operational centres and delegate offices in 19 countries. Its international offices, based in Brussels and Geneva, are responsible for liaising with international organisations. *World in Crisis*, published as part of MSF's annual International Day for Populations in Danger, is an integral part of MSF's commitment to the independent provision of humanitarian aid to the victims of conflict and disaster throughout the world.

WORLD IN CRISIS

*The politics of survival at
the end of the twentieth century*

•

MÉDECINS SANS FRONTIÈRES
DOCTORS WITHOUT BORDERS

Edited by

Médecins Sans Frontières
Doctors Without Borders

MSF project coordinator: Julia Groenewold

Associate editor: Eve Porter

LONDON AND NEW YORK

First published 1997
by Routledge
2 Park Square, Milton Park, Abingdon, Oxon, OX14 4RN

Simultaneously published in the USA and Canada
by Routledge
711 Third Avenue, New York, NY 10017

Transferred to Digital Printing 2008

Typeset in Garamond by Florencetype Ltd, Stoodleigh, Devon

British Library Cataloguing in Publication Data

A catalogue record for this book is available from the British Library

Library of Congress Cataloguing in Publication Data
A catalogue record for this book has been requested

ISBN 0–415–15378–6

Publisher's Note
The publisher has gone to great lengths to ensure
the quality of this reprint but points out that some
imperfections in the original may be apparent.

CONTENTS

•

PREFACE

Doris Schopper

•

On its International Day for Populations in Danger, Médecins Sans
Frontières tries once more to draw the attention of the public to the difficulties, challenges and dilemmas of providing humanitarian aid to civilians.
Our last report on populations in danger, which was published in the first
quarter of 1995, was in great part devoted to the genocide in Rwanda, its
origins and its consequences. It sought to draw attention to the failure of
the international community to intervene early and decisively enough to
prevent the systematic and well-planned massacres from happening.

Since this first emergency, we have also questioned the massive provision of humanitarian assistance to the refugee camps in Zaire and Tanzania.
We have witnessed the transformation of these camps into a base from
which Hutu extremists, guilty of participation in the genocide, plan
revenge, thus making reconciliation between the two ethnic groups an ever
more difficult undertaking. Violence breeds violence: the massacre of several
thousand Hutus during the evacuation of a camp in Kibeho by the Rwandan
military in April 1995 is part of a vicious circle that humanitarian aid
cannot break and sometimes may even fuel.

The role and difficulties of humanitarian organisations in situations such
as Rwanda, but also in Burundi and more generally in east and central
Africa's Great Lakes region, in Liberia, in Sudan, in Chechnya and in Bosnia
will be explored in various ways in this book. How can the needs of civilians be balanced against the negative consequences of the assistance
provided, particularly when the abuse of humanitarian aid by one or several
of the conflicting parties fuels the conflict, as has happened in the camps
around Rwanda and has proved to be the case in Liberia?

How is it possible to choose between being a silent presence helping the
victims and speaking out about unacceptable human rights abuses at the
risk of expulsion? This is a dilemma which we have faced and which we
still face most crucially in the Great Lakes region, but also in Sudan and
in Chechnya. What is the role of the United Nations protection forces in
such situations as those in and around Rwanda? Or in Bosnia, where we
have witnessed their presence dramatically misleading refugees and displaced

persons by giving an impression of ensuring a security which in reality did not exist? This was certainly the case in Kibeho, but also in Srebrenica and Zepa. As Rony Brauman points out in his Foreword, the ones who wanted to provide protection were not in a position to do so, and the ones who were in a position to provide protection refused.

The issue of how to protect civilians in conflicts which not only deliberately target civilians, but use them as hostages and human shields, is thus of the essence in the current humanitarian environment. This book tries to address some of the most burning questions.

Who can guarantee protection? What is the role of humanitarian organisations? Who should intervene when humanitarian assistance can no longer be provided for civilian victims of conflict? However, beyond the issue of protection in zones of conflict, we are also faced with an increasing problem of access to safe areas for people who are persecuted on political, ethnic, religious or other grounds. Over these past few years, the global growth of violence and conflict has swelled the world's refugees and displaced populations to over 50 million people. The international system for protection of refugees established after the Second World War is being challenged by growing demands for asylum and by a change in attitude among many Western governments. Prior to the fall of the Berlin Wall, refugees were often considered as welcome symbols of the failure of the communist system. Today, they are considered economic migrants and as an extra burden in a difficult social and economic climate.

There is great concern that refugees are no longer granted the right to protection and asylum, and that forced repatriation and even rejection may become acceptable solutions. The situation is even more dramatic for the internally displaced persons, who are not recognised as traditional refugees because they have failed to cross international borders. These people are even more vulnerable, as they have neither international protection nor the effective protection of their own governments.

The 1951 Convention on the status of refugees does not apply to displaced persons. We are currently witnessing a particularly dramatic example of this type of situation in the Kivu region of Zaire, where the status of the Banyarwandans, is uncertain and, depending on the observer, varies from refugee to displaced person to stateless. The recent influx of Rwandan refugees has exacerbated this problem, thereby leaving more than 250,000 people without official protection in a life-threatening situation.

It thus seems that we are witnessing a widening gap between the proclaimed intention to protect civilians who are victims of persecution and war, and everyday reality. However, international humanitarian law, which was first developed more than a century ago and further improved in the past decades, should enable the international community to address some of these issues. It should help determine the responsibilities of individuals and organisations and, when necessary, impose sanctions.

In recent years, the difficulty of applying humanitarian law has been demonstrated in many circumstances. In the past year alone, it was dramatically evident in Rwanda and Bosnia, where the perpetrators of genocide and mass ethnic cleansing are still free and continue to wield power over important parts of the population, with no sign of remorse. The establishment of two tribunals to prosecute war criminals from the former Yugoslavia and Rwanda are a step in the right direction, but they still need to prove their ability to convict those responsible.

Beyond humanitarian law, the Universal Declaration of Human Rights, framed in the aftermath of the Second World War, provides us with an even more powerful tool to challenge those governments which abuse individuals and groups. But putting it into practice is much more difficult. The UN human rights bodies work slowly and often inefficiently: this comes as no great surprise. How can an organisation that is directly dependent on governments accuse one of its own members and intervene in what is largely considered an internal affair? Human rights monitors such as those deployed in Rwanda may prove useful (although this remains to be proved) if continuity were ensured through long-term funding, and if their role were properly clarified: are they there to observe or to protect? It is clear that existing legal instruments have their shortcomings, but if there was truly the political will to apply them, much could be improved.

Given the increasing complexity of providing humanitarian assistance and the risk of it being used as a political instrument, it is all the more crucial that humanitarian organisations remain totally independent of all political interference by governments. We should not accept our actions being used as a political fig-leaf. The international community at large, and western governments in particular, have to accept that it is not the role of humanitarian organisations to provide the solutions to a problem. Our role is to help the people in danger survive beyond the crisis and to pressure governments to assume greater political responsibility in searching for short and longer term solutions.

The rapidly changing political and economic climate at the end of this century will certainly influence our perception of the role humanitarian organisations should play in the world of tomorrow. Since MSF was created 25 years ago, the humanitarian environment has greatly changed. The East–West divide has ended and conflicts are no longer determined by allegiance to the capitalist or communist camp. As nations established by force fall apart, civil wars erupt that will leave deep physical and also psychological scars for decades to come. And as the economic climate becomes more tense, the number of people stricken by poverty and/or discriminated against (including ethnic minorities) grows in western societies as well as in the former Soviet Union.

Diseases which seemed to belong only to the Third World have reappeared in New York and Moscow. In addition, the AIDS epidemic has spared no country, poor or rich. It will accompany us into the twenty-first century, bringing in its train exclusion, rejection and suffering. In about 20 years' time, 60 per cent of the world's population will live in urban areas, increasing the number of people living under difficult conditions in under-serviced townships and breeding even greater poverty and violence. What should the role of humanitarian organisations be in dealing with such situations?

What is the role of governments? Should we become providers of services on an international scale, fighting the negative effects of an economic crisis, or should we denounce the growing egocentrism in our societies and put pressure on governments to take action?

Within MSF, the debate about 'our' role is lively and ongoing, and in several instances this book provides some critical views. Since its creation 25 years ago, the action of MSF has always inseparably combined provision of medical aid with personal commitment to act as witness to the plight of populations in danger.

At the end of the second millennium the world has on the one hand become a global village connected by fax, phone and Internet, where anything that happens is immediately brought to the attention of millions of people, where we may have the impression that we can do anything and be anywhere in a matter of hours or, at the most, days. On the other hand, as we are overwhelmed by the sheer amount of information, as we are unable to cope with so much suffering and violence and as we feel vulnerable in an environment we cannot control, our societies close in, we become concerned solely with our own backyards and try to forget that as human beings we bear responsibility for each other. In this context, the responsibility to witness has an essential role to play, not only to inform about people whose very survival may be threatened, but to show how they can be helped to survive and to regain their human dignity.

PLATES

PLATES

MAPS

CONTRIBUTORS

Françoise Bouchet-Saulnier   Françoise Bouchet-Saulnier, a doctor of law specialising in international law, human rights and humanitarian law, has headed MSF's humanitarian law section since 1991. She has extensive experience working in and consulting on international law for NGOs, governments and international organisations. She is currently a Research Director at the MSF foundation in Paris.

Rony Brauman   Rony Brauman was chairman of MSF-France from 1982 to 1994. A medical doctor, he has worked in international medical assistance since 1977 and is author of several articles and books on topics relating to crisis relief, human rights and humanitarian issues. He is currently Research Director at the MSF foundation in Paris.

Vincent Faber   Vincent Faber has worked as a geophysicist and as a management consultant before becoming involved in humanitarian work in 1991. Having worked for Médecins du Monde for several years, first as a field worker in many different areas and then as a programme manager in the Paris office, he joined Médecins Sans Frontières (Holland) in 1995. Until April 1996, he was country manager in Zaire, based in Goma.

Pierre Gassmann, a Swiss citizen, is currently head of the ICRC in Columbia. He first joined the ICRC as delegate in Biafra at the age of 22. Since then he has represented the ICRC on numerous assignments both at headquarters in Geneva and abroad, including as Delegate General for Africa and senior adviser to ICRC president Sommargua.

Julia Groenewold   Julia Groenewold, *World in Crisis* project coordinator, has worked with MSF since 1989. She is currently working in the Communications Department, coordinating activities for MSF's International Day for Populations in Danger.

Iain Guest   Iain Guest is a British journalist based in Washington, DC. He has reported world-wide for the *Guardian*, *International Herald Tribune*,

BBC and others on humanitarian, environmental and human rights issues. He is also the author of several books.

Marelle Hart   Marelle Hart works in the project department of MSF-Holland. She has worked as a radio journalist for Radio Netherlands World Service and the BBC World Service.

François Jean   François Jean is currently Director of Research at the MSF Foundation and has worked with MSF since 1982. He is author and editor of several articles and books on famine, conflicts and humanitarian action, including publications for MSF.

Ed Schenkenberg van Mierop   Ed Schenkenberg van Mierop serves as MSF-Holland's policy adviser on humanitarian affairs. He studied public international law at Leyden University.

Stephan Oberreit   Stephan Oberreit is an economist who has been with MSF since 1993. He has administered programmes in southern Sudan and he has also been a coordinator in the former Yugoslavia. He is currently development director of the MSF foundation.

Stephan van Praet   Stephan van Praet, a sociologist and anthropologist, currently heads the research centre at MSF-Brussels. He has worked with MSF since 1986, mainly in West Africa, Sudan and the Horn of Africa.

Pierre Salignon   Pierre Salignon is a lawyer and a former coordinator of MSF programmes in Croatia. He is currently a deputy desk officer for MSF in Paris.

Doris Schopper   President of MSF-Switzerland since 1991 and President of the International Council of MSF (February 1995 to June 1996), Doris Schopper is a medical doctor who has worked with MSF since 1982. She has also served with the World Health Organisation and as a lecturer in public health.

Renaud Tockert   Renaud Tockert is a doctor who has worked for MSF-Belgium since 1987. He spent three years as regional coordinator in Mozambique and then became a member of the MSF-Belgium emergency team, spending most of his time in Iraq during the Gulf crisis. From 1991 to 1995 he was desk manager of the project department responsible for (among others) Iraq, Bosnia, Afghanistan, Mozambique and South Africa. He is now head of the MSF-Belgium PR department.

Mike Toole   Michael J. Toole, a medical epidemiologist, is head of the International Health Unit at the Macfarlane Burnet Medical Research Centre in Melbourne, Australia. He has worked in international medical relief since

1971 in positions involving extensive work in the field, including rural and refugee public health, establishing health programmes in developing countries, and coordinating the US Centers for Disease Control and Prevention technical assistance programmes to refugees and displaced populations. He is a founding board member of MSF-Australia.

Fabrice Weissman　Fabrice Weissman is in charge of research at the Médecins Sans Frontières Foundation in Paris. He is author of a number of studies and articles on conflicts in Africa.

FOREWORD

The dawn of the twentieth century brought genocide against the Armenians. It happened in near-secret, with no media or humanitarian organisations to tell the world what was going on. The close of the century brought genocide against the Tutsis. It happened at a time when humanitarian concerns had a higher profile than ever before, in a country which had a contingent of 2,500 United Nations Blue Helmets, and where dozens of international non-governmental organisations (NGOs) and the press were present.

Teams from Médecins sans Frontières (MSF) were hard at work providing relief to the hundreds of thousands of displaced persons who had been forced into the north by the civil war, and in the Burundian refugee camps in the border provinces of the south. But we were no more aware than anyone else of what was to come. Nevertheless, like everyone else, we *did* know that an outbreak of violence was threatening. Being there, on the ground, is an excellent way of gauging human suffering. Yet it does not necessarily result in understanding the causes of that suffering or how to prevent it.

In April 1995, three thousand Hutus living in a camp in Kibeho, the former 'security zone' set up by the French army in the west of Rwanda, were slaughtered by the Rwandan army. This occurred under the noses of the MSF staff and a detachment of Blue Helmets. The relief workers were able to tell journalists what they had seen with their own eyes, but only after the event, when the damage had been done. They had at least spoken out. The UN soldiers, on the other hand, had received no orders to take action. Just as they had during the massacres of April 1994. As always, the ones who wanted to provide protection were not in a position to, and the ones who were in a position to provide protection refused. The ones whose *duty* it was to provide protection, the rulers of Rwanda, were at best in cahoots with, and at worst actually giving orders to, the perpetrators of the massacre, just as had been the case during the army massacres in the months after the Patriotic Front seized power.

DEADLY ILLUSIONS

That nobody was able to offer the slightest protection against the events in Kibeho is not particularly unusual. The question that arises, however, and there are certain parallels here with the massacres following the fall of the Bosnian enclave of Srebrenica, is whether the presence of UN soldiers had actually made things worse. Did their presence give the refugees the deadly illusion that they would be protected when in fact they would not? The humanitarian organisations providing relief in such instances are only there to do their job. At best, they can hope only to deter attack by acting as witnesses. The armed UN soldiers have a completely different, and deceptive, image. They lend the impression that they are able to provide authority in places where there is none. Their presence suggests an international legitimacy capable of shielding people from danger. News in our global village, however, travels far more slowly than we might like to think. Many people still believe that sending in the Blue Helmets is a sign that something is being done. Sooner or later, they will realise that such operations are really a sophisticated way of signalling that nothing, in fact, will be done. All that these troops are providing is a form of telematic window-dressing.

It is a common misconception that the deliberate slaughter of civilians is something which arrived with the awakening of identities provoked by the end of the Cold War. Bloodbaths are not simply recent phenomena. The twentieth century is full of them. What is new, however, is the presence of humanitarian organisations on the ground. These are international entities, notably NGOs and UN relief agencies, with no ostensible government ties. They are being given implicitly, and in some cases *want* to be given, a part in protecting civilians. In other words, they place themselves in one of the fundamental roles of a government: that of ensuring the collective safety of its populace.

Apparently dizzied by its media successes, the humanitarian movement has progressed in a few years from assuming an ethical responsibility, which it has rarely succeeded in handling effectively, to taking on a legal responsibility that has proved impossible to deal with. In the case of MSF, the concept of ethical responsibility is one which has come far in the organisation's 25 years of existence. Yet it has always aroused heated debate between and within the various country-based sections.

In its early days, MSF saw itself first and foremost as a watchdog. The first generation of volunteers, who lacked the material resources to organise genuinely efficient medical aid, sought to act as 'runners.' They brought information to the Red Cross and UN agencies, whose own activities were slowed down by the legal and diplomatic procedures which bound them. They also sought to use the media's ability to mobilise public opinion and therefore exert pressure on institutions as an essential tool for promoting humanitarian activities.

THE SALUTARY MISTAKE OF BIAFRA

The job here was not to protect, but to speak out. But what did speaking out mean? Did it mean confining ourselves, as the neutrality of humanitarianism dictated, to describing suffering, or could we talk about its causes, thus exposing ourselves to accusations of taking sides? Did it mean having a scale of events and drawing a demarcation line, or did it mean reporting every example of distress, at the risk of cataloguing every form of human madness?

There was nothing theoretical about these issues for the founders of MSF. They stared them in the face during their mission in Biafra in 1968 and 1969. Like the Christian missionaries and Oxfam, they decided to side with the Biafran secessionists, and echoed the accusations of genocide which the separatist Ibo leader General Ojukwu was levelling at the Nigerian government. They decided to keep quiet about what they had seen in the Biafrans' stronghold: a starving mass kept that way by Ojukwu to show the world's press. Convinced that they were witnessing the annihilation of a people and refusing to imitate the guilty silence observed by the Red Cross during the Second World War, they had sided with the victims, that is, with the Biafrans.

Suffering is no guarantee of political or moral truthfulness, however. General Ojukwu would not countenance a road corridor through Nigerian territory to resupply Biafra. His intransigence helped to reinforce the blockade and increase the starvation. The ideal of Biafra was far more important in his eyes than the lives of its inhabitants. The Nigerian federal government's goal was not to destroy the Biafran people but to keep the country together. Apart from isolated cases of brutality, the army did not engage in reprisals after it had won in January 1970. This error of judgement, although entirely understandable in the circumstances under which the relief workers were operating in Biafra, created a salutary rift in the humanitarian movement's tradition of neutrality.

This rift is one which has been alternately patched over and then reopened throughout the years of MSF's existence, a sign of the difficulty of reconciling two legitimate desires. The first is the desire to speak out loud and clear to avoid siding passively with an oppressive overlord, or simply to expose a situation of which the media are unaware. The second is to keep the overlord sweet in order to continue to be able to help the victims on the ground, which is after all the aim of any humanitarian organisation. Over the years, however, experience and reflection have gradually brought these issues out into the open. MSF's attitudes were shaped in Afghanistan, invaded by the Soviet Union, and in Cambodia, occupied by Vietnam, before being put to the test in Ethiopia.

COPING WITH COMMUNISM

In Cambodia, MSF found itself on a collision course with the pro-Vietnamese powers in Phnom Penh. Refusing to submit to the government's desire to exercise total control over international aid, MSF's leaders (including the author) organised the extremely controversial March for the Survival of Cambodia. It gathered on 6 February 1980 on the Thai side of the Cambodia–Thailand border. The associations involved fully realised that they were kissing goodbye to any prospect of being able to operate in the country after their act of provocation, but they wanted publicly to unmask the predatory dictatorship which had replaced a rule of terror. The aim was to show the international community that aid was being entirely controlled by a pitiless government which was using it to strengthen its stranglehold over the population. A fierce media battle ensued between the relief organisations. It pitted those which credited the Cambodian government with genuine concern for its people against those which considered it tyrannical and corrupt. The intrusion of politics into humanitarian activities can be seen to date from before the fall of the Berlin Wall.

In Afghanistan, there was never any question as to whether MSF should offer its services to Kabul in order to be able to sit on the fence. Medical cover was sufficient in the government-controlled urban areas. There was no dilemma. As with the vast majority of humanitarian organisations active in Afghanistan, MSF never sought to take a neutral stance. We contrasted the legitimacy of our one-sided operations in the camps of the resistance with the legality of being on both sides. Together with the other aid organisations, we told the world about the scale of this war in which a million died.

Our aid unapologetically involved publicly denouncing the atrocities committed by the occupier, backing the investigations into its war crimes and speaking out to official bodies like the United States Congress and the European Parliament. Like our forerunners in Biafra, we had implicitly picked our side. We had made our choice clear by undermining, deliberately and indirectly, the diplomatic position of our side's enemy. We all saw it as our duty to expose the scale of this war to the world, especially since it was mentioned so seldom in the media during its early years.

It is worth pausing to reflect on the fact that MSF's philosophy of speaking out was largely born of its regular confrontation with communist regimes. We have been in most of the world's refugee camps, where fully 90 per cent of their inhabitants were people fleeing communist rule. It was the result of internal strife provoked by the communists and their use of brutality that caused this exodus to swell the camps. Resistance strongholds and refugee camps gave other people's troubles a face: that of Soviet imperialism. This explains why the principles which MSF defended in the 1980s owed more to French writers Albert Camus and Raymond Aron than to those who theorised about the Third World. Our battle was to inform public opinion about the brutality of totalitarian regimes to civilians. We sought to prevent these regimes from misappropriating aid and to defend the independence of humanitarian action. While we are on the subject, it needs to be pointed out that when Bangladesh expelled over 200,000 Rohyinga refugees to Burma in 1978, MSF, which was working in the camps, did not protest. Nor, for that matter, did anyone else. Before the exodus of the boat people and the outpouring of support for them, nobody had questioned the right of a country to turn away refugees.

KILLER AID

Colonel Mengistu's Ethiopia, yet another communist country, was to be the scene of a direct confrontation between MSF and a government over humanitarian principles. We all remember the famine of 1984–5 and the huge aid effort which was sparked by the BBC television report of October 1984. We might also remember how 800,000 people were forced to migrate from the north to the south in 1985, and how 200,000 of them died in the process. We may well have forgotten, however, how international aid was used to back up this brutality, if only by playing on the trust which the presence of foreign volunteers in the camps inspired in the people who came to live in them. The same people would then be rounded up by the militia and the party.

The non-governmental organisations (NGOs) provided the bait. The feeding centres had become traps. The trucks and cash given by the international community were requisitioned to widen and speed up the forced migration campaign. The Ethiopian regime used the silence of the NGOs to dispel the fears which were expressed from time to time. Their silence was read as approval. As had occurred in the Second World War, the neutrality of the humanitarian organisations provided a benign facade for totalitarian power. The silence succeeded only in aggravating the brutality. MSF failed in its bid to persuade the NGOs to rally round in a joint protest. Isolated in its solitary refusal to endorse this deadly practice, it was expelled from Ethiopia in December 1985. A few months later, however, pressure from the biggest donors, the European Community and the USA, brought a halt to the deportations. MSF's determined campaign played no small part in this development, although it was admittedly also driven by the Reaganite thinking of the time and the gathering collapse of communism.

As with Cambodia, MSF refused to be a party to this humanitarian version of building Potemkin villages or to blinding people with good intentions. When aid turns its back on its aims and aids and abets the oppressor while further cowing the victims, the first duty of a humanitarian organisation is to try and prevent the abuse of compassion that is going on, even if it risks its own neck in so doing. It cannot be denied that most of the humanitarian community opted to let things continue and to cope with the issue of responsibility by ignoring it. In such critical situations, the humanitarian organisations have favoured the indirect approach in order to serve the needs of an overriding, but not often recognised, objective: not causing any harm. A wider look at humanitarian activities reveals that their only direct power to protect comes from the deterrent effect of an international presence. There is evidence to suggest that this deterrent effect is not worth much, and that in some (fortunately rare) cases, it can in fact make things even worse for those supposedly being helped. This happens through silence. It is not the presence itself which does the damage, but the use of that presence for propaganda purposes. This is, after all, the accusation often levelled at the Red Cross, which kept its mouth shut about Auschwitz. But it is much more convenient to consign this episode to the past than to ponder similar failures to speak out and the reproduction of such types of blindness.

It has been 12 years since the Ethiopian famine, and the fall of communism has seen events unfold at an incredible pace. The world's alliances and balances of force have changed. Europe has its own flashpoints of fighting. Conflict has produced fragmentation and warring parties are multiplying. More than ever, humanitarian concerns are in fashion, attracting western countries in search of new images and prospects. State-sponsored humanitarianism and the new role of the UN have further blurred the already fuzzy identity of the humanitarian action. With or without Blue Helmets in Somalia or Zaire, the new world order has done nothing to change the issue of protecting civilians.

STABILITY VERSUS PROTECTION

It is in fact the international institutions of the twentieth century, the League of Nations and later the UN, that put an end to the concept of international protection by outlawing the use of force in relations between countries. Russia initiated humanitarian intervention at the end of the eighteenth century when it sought to protect the Orthodox subjects of the Ottoman Empire. This right was invoked a number of times during the nineteenth century too. France used it to defend Maronite Christians under attack from the Druze, a situation which ended with autonomy for Lebanon. It was lawful to use force to 'end practices considered shocking to the human conscience'. This is what India did in East Pakistan in 1971, when its armed intervention resulted in the birth of Bangladesh. It is what Vietnam did in the Cambodia of the Khmers Rouges in 1979. And it is what Tanzania did in Idi Amin's Uganda, also in 1979. Some countries considered such an imperial style of behaviour to be unacceptable interference, but some peoples considered it a ray of hope that would gain international recognition. Such actions are not quite a thing of the past. This was made clear by Operation Provide Comfort, aimed at assisting Iraq's Kurds in the wake of the Gulf War. Normally, it is the threat to international peace and security, as in Somalia and former Yugoslavia, that is used to justify the deployment of force. The overriding concern, as was also the case in Rwanda, was to halt movements of refugees which were threatening to spill over into neighbouring countries. Isn't the phenomenon of what some consider to be the beginning of a new world order for humanitarian activities in fact only a modern manifestation of the obsession with stability which has haunted governments since the First World War? The answer lies in the question itself. Yet this obsession with stability should not necessarily be condemned, given that it is fully justified by events.

The fact is, the international community has never before shown such duplicity, nor has the gulf between words and deeds ever been so wide. The atrocities of the Khmers Rouges, whose cruelty compares with that of the Nazis, never prevented the international community from talking to them and granting them a political recognition which they still have. The Russian army's massacre of the Chechens has not prevented the West from rushing to support financially and diplomatically Boris Yeltsin, who is the main instigator of the war in Chechnya. Slobodan Milosevic, the chief architect of the war in former Yugoslavia, is one of the sponsors of the Dayton peace accords, which contain clauses on the return of refugees which are being violated daily in full view of the very people charged with enforcing them. And the list of examples of cynicism, expediency and blindness goes on. However we describe them, the underlying fact is the same: a huge gulf between words and deeds.

Rather than complaining, the NGOs should seek to use this obvious discrepancy to further their humanitarian aims. They should relentlessly remind governments of their statements and commitments and blow the whistle when countries go back on their promises.

MACHINES AND PRINCIPLES

This alone, however, would not be enough. In some circumstances, given the development of humanitarian assistance, its dramatisation, its political use and some of its negative consequences over the past two decades, we need to go further than blowing the whistle.

The use and abuse of the presence of NGOs and their operations can sometimes cause a lot of damage. NGOs then acquire a responsibility, a fact which they all too often fail to recognise. This is probably because it is so difficult to countenance and accept that something being done with the best of intentions is in fact harming those it is supposed to be helping. Another major reason is that the interests of the NGOs themselves can be seriously jeopardised by conflict with political authorities. Humanitarian organisations have become multinational machines with big budgets: they have a lot of people to manage and have been sucked into a spirit of technicism and moralising conformity which is making them look more and more like the UN.

The logic of self-preservation is all too easy to confuse with an activism which is blind to its own consequences. The worst, but not the only, examples of this thinking have been seen in Ethiopia, Bosnia and Rwanda. It can be seen when NGOs talk about protection. They should look not just at the countries of the world and international organisations, but at themselves. They might ponder the role that they have so easily and eagerly slipped into (one that implicitly has been recognised as theirs) in recent years, notably that of spokesman for those with no voice and no rights, a sort of victims' trade union.

This self-examination has been seen from time to time in international humanitarian circles. For MSF, which is one of the first in the firing line of criticism, one of the forms of self-examination is the annual *Populations in Danger* report. I sincerely hope that this, the fourth in the series, contributes to this vital exercise.

Rony Brauman

MSF MISSIONS THROUGHOUT THE WORLD 1995–1996

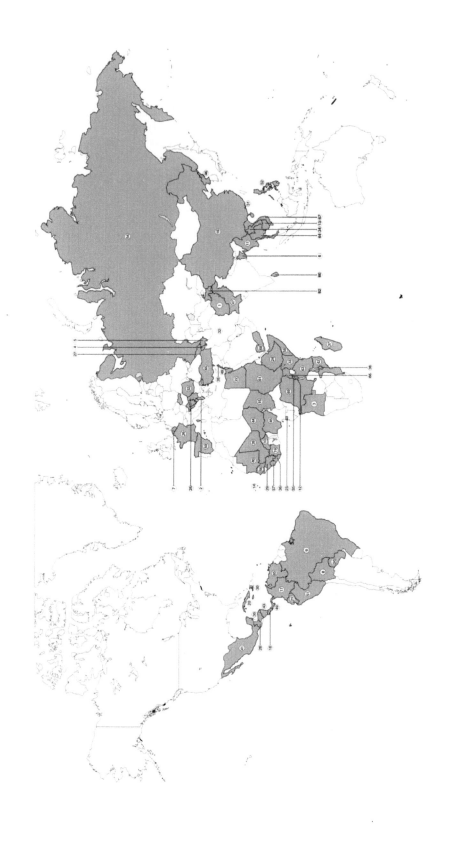

MSF missions throughout the world 1995–1996

1	Afghanistan	19	Cote d'Ivoire	37	Madagascar	55	Rwanda
2	Albania	20	Cuba	38	Malawi	56	Senegal
3	Angola	21	Ecuador	39	Mali	57	Sierra Leone
4	Armenia	22	Egypt	40	Mauritania	58	Somalia
5	Azerbaijan	23	Equatorial Guinea	41	Mexico	59	Spain
6	Bangladesh	24	Ethiopia	42	Mozambique	60	Sri Lanka
7	Belgium	25	Bosnia/Croatia/FYR	43	Nicaragua	61	Sudan
8	Bolivia	26	France	44	Niger	62	Tajikistan
9	Brazil	27	Georgia	45	Nigeria	63	Tanzania
10	Burkina Faso	28	Guatemala	46	North Korea	64	Thailand
11	Burma/Myanmar	29	Guinea	47	Pakistan	65	Turkey
12	Burundi	30	Haiti	48	Palestinian Authority	66	Uganda
13	Cambodia	31	Hong Kong	49	Panama	67	Vietnam
14	Cape Verde	32	Iran	50	Paraguay	68	Yemen
15	Chad	33	Kenya	51	Peru	69	Zaire
16	China	34	Laos	52	Philippines		
17	Colombia	35	Lebanon	53	Romania		
18	Costa Rica	36	Liberia	54	Russian Federation		

PROTECTION OF CIVILIANS
IN CONFLICT

Ed Schenkenberg van Mierop

•

Civilians have always been under threat in war. But the methods of modern warfare seem sometimes to threaten more of them more of the time. In recent years wars have seemed characterised by endless streams of wretched refugees, fleeing violence or mayhem or starvation, being corralled into camps. The western agencies attempt to dispense both aid and protection, often in competition with one another and with the political or military authorities which attempt to dominate the refugees.

Many of these agencies do their best under horrific conditions – and sometimes their best is good. But all too often the help they can offer is at best a short-term palliative. Their works, and the charitable inclinations of their supporters, are exploited for political ends. They are often forced to be part of the problem, not part of the solution. Kurdistan, Somalia, Bosnia, Rwanda, Afghanistan, Liberia – the list goes on and on, with the images of those suffering blurring in and out of focus as attention flickers from place to place, disaster to disaster. At the end of the day the impression that lingers, rightly or wrongly, is that all over the world people are inadequately protected from violence. It is not new that civilian casualties are the purpose rather than the by-product of war. But the numbers of victims are increasing exponentially, and responses seem more and more inadequate.

I suppose one can argue that in the last fifty years there have been three different periods, three different kinds of warfare. First, during and just after the Second World War, conflict was classical – states fighting each other. Second, during the Cold War, the superpowers dominated all. In the period of decolonisation, governments fought guerrilla liberation movements, which were often based on desire for independence and on some form of political morality. Humanitarian organisations had to find some way of working with these movements. The Cold War established certain patterns. In most conflicts there was polarisation induced by the bipolar world. So there was a framework in which the parties were identified.

Since the end of the Cold War, established patterns are now gone. Now there is often a surplus of parties or interlocutors. But none has real authority. This new, third period started after the fall of the Wall in 1989; it is a period of non-structured or destructured conflict – as in Somalia, Liberia, Rwanda, Bosnia – which is sometimes called identity-based. In Somalia, a senior official of the International Committee of the Red Cross (ICRC) once said that almost every individual was looking for his own identity and that was the main motivation for the conflict.

Certainly, one can agree with Eric Morris of the United Nations High Commissioner for Refugees (UNHCR), who argues in a new publication that what is happening now is that for the third time this century the world is in the throes of the creation of new states emerging from the collapse of an old world order. And with this process of creation the demands of and for ethnicity have become ever more virulent. No one quite knows what it means – ancient hatred, as Morris points out, has become the catchall cliché to describe and explain it – but its effects have been powerfully destructive, from Rwanda to Bosnia and within the former Soviet Union. Ethnic demands have played an important part in creating something approaching chaos in international relations – at the least disorder.

Since the fall of the Wall, that amorphous entity known as the international community – which usually means a few developed and articulate nations led by or at least influenced by the United States – has made a distinction amongst conflicts. It has been between those which are seen as a real threat to the world's vital interests – e.g. the invasion of Kuwait – and those that are not – e.g. Rwanda, Somalia and Bosnia. In the first case, peace will be enforced effectively, indeed ruthlessly; in the second, negotiations, monitoring and humanitarianism are deemed adequate. One writer has described this approach – the predominant approach – as 'containment with charity'.

The idea of 'humanitarianism' itself has been inflated. The most obvious and most telling recent example of this has been Bosnia. The West never defined a political objective for the former Yugoslavia. Humanitarianism was used as a cloak for this failure. The badly named UN Protection Force, UNPROFOR, was established not to protect the citizens of former Yugoslavia but the humanitarian relief programmes run (efficiently) by the UNHCR. UNPROFOR undoubtedly saved lives and alleviated much misery. But its other effect was – as with so many relief operations – to reinforce the war parties and extend the war.

Almost all humanitarian agencies try to claim that they are neutral and impartial. This can often be a curse. What matters is the perception of them by different actors in a conflict. In Bosnia the UN and its agencies, supposedly neutral, were decried as biased by all sides. The ICRC rigidly refused to accept the protection of UN armed escorts in former Yugoslavia and ICRC officials argued that they were therefore more widely accepted as truly neutral.

The sorrow of Domanovici

In 1993, I worked as a medical officer with MSF-Holland on a drug distribution programme in war-torn Hercegovina. Besides supporting the existing medical infrastructure there, we assessed the needs of vulnerable groups like displaced persons and psychiatric patients. During this assessment we came across a mental hospital called Domanovici.

The building had been partly destroyed by the Yugoslav People's Army (JNA) at the beginning of the war. Though it had been recently repaired by Swiss Disaster Relief, a reconstruction NGO, we found the hospital completely deserted. Its 212 patients were evacuated in 1992 because of nearby shelling and fighting and were sent to four different buildings on the West Bank in Mostar, which was a safe place at that time.

During the spring of 1993 a very bloody war broke out between the Bosnian Government Forces (BiH) and the Bosnian Croats (HVO). We visited one of the buildings where the psychiatric patients lived and saw that the living conditions were appalling. The patients wore only rags and suffered from scabies, lice and fleas. The food was of poor quality and medical care was virtually non-existent. Their fourth-floor accommodation was accessible only by a staircase left completely exposed by shelling.

But what worried us most was the security situation. The building was only 100 metres away from a very active frontline, and the patients were easy targets. Because of their mental disorders they were completely oblivious to security, and all of them were heavily sedated with strong neuroleptics, which caused them to move around very slowly.

When we explained to the UNHCR Protection Officers in Medugorje that it was a serious violation of these patients' human rights to house them so close to the frontline, the officers agreed to help evacuate them back to the Domanovici mental hospital. But the HVO didn't allow them to go to Mostar's West Bank. When we then discussed the situation with the Minister of Health of the self-proclaimed Croatian Republic of Herceg-Bosna, he immediately said that he was not in a position to change anything.

In the meantime, seven patients were murdered and nine wounded by snipers. Five patients who could no longer stand the suffering committed suicide by jumping out of windows. One patient was killed by a grenade while begging on the street for cigarettes. Two others were killed by their own police while unknowingly violating the curfew.

Many innocent people died for nothing. UNHCR knew what was happening, but was powerless to do anything and – even worse – accepted its powerless role. Local high-ranking politicians knew exactly what was going on. But they had other interests and closed their eyes.

The Croatian authorities, we learned, had a hidden agenda. They did not want the psychiatric patients back in Domanovici because they had already designated the building to accommodate Croatian displaced persons. Even worse, HVO wanted to use the building for military purposes.

Our only recourse was to inform the public by releasing the story to the media, including the BBC. Following the Dayton accords, some of the patients were evacuated to Croatia, others to Italy. They never went back to Domanovici.

Fokko de Vries, Dutch Medical Officer, MSF, Mostar

Plate 1.1
Refugees at a
cholera treatment
centre in Goma,
Zaire. Photograph:
Teun Voeten

Indeed, so-called 'neutral' and 'impartial' intervention can and often does have exactly the opposite effect from that intended – it exacerbates the suffering by prolonging the conflict. All this behind what has been called the humanitarian alibi or even 'fig-leaf'. But those attempting to use the alibi and to hide behind the fig-leaf have found that in recent years many humanitarian operations have become more difficult. Interlocutors cannot be relied on. And funding for protection is harder and harder to find. Indeed, it would be impossible alone. Protection has to be carried and financed on the back of relief operations.

Another change is that previously the parties – guerrillas and governments – received a lot of assistance from outside. Now they need to raise money. But here a conflict arises. Relief requires the cooperation of the authorities – as Bosnia demonstrates. Protection often demands confrontation. In Bosnia and elsewhere the traditional duty of protection of individuals has often been sacrificed to the logistical requirements of the feeding systems. More time has been spent on protecting wheat and rice than people.

In Africa there have been, in the last two years, at least 25 countries in which civilians have been subjected to greater or lesser degrees of violence. Amongst those places where there have been the most serious abuses are the Great Lakes region (East Zaire, Burundi, Rwanda); Angola; Liberia;

Sierra Leone and neighbouring countries; southern Sudan; Somalia – because no real solution has been found there.

In Liberia the ICRC was forced to evacuate its foreign staff in April 1996 – for the fourth time in the country's six-year war – because conditions had once again become so dangerous. Jean-Daniel Tauxe, a senior ICRC official, wrote chillingly, 'Teenage fighters high on drink or drugs steal our vehicles. They then drive off to bring in reinforcements who follow suit, multiplying the looting and anarchy at the expense of the ordinary Liberians we seek to help. What do we do?' he asked. His answer was that humanitarian organisations could do nothing – the solution had to be found in the Security Council.

In Somalia, the humanitarian operation (UNOSOM) failed because no one understood what the troops were there for. UNOSOM failed also because soldiers are now no longer allowed to do their jobs. A principal reason why UN operations falter today is the zero casualties option. One of the biggest misunderstandings at the end of this century is that people expect military–humanitarian missions to bring peace and stability. As presently arranged, they cannot. Indeed, it is arguable that if soldiers are not prepared (or allowed) to do their traditional job they are rarely worth deploying.

Plate 1.2
Burmese Nobel
Peace Prize laureate
Aug San Sun Kyi
speaks to a crowd
in Rangoon, Burma.
Photograph:
Jan Banning

Plate 1.3
Armed mujahideen, or holy warriors, formerly involved in Afghanistan's decade-long war against the Soviets, are now caught up in a civil war of their own with devastating costs for the country's civilian population.
Photograph: Bart Eijgenhuijsen

The lessons of Somalia, learned in particular in Washington, were applied to most awful effect in Rwanda – so far. It was here in 1994 that genocide, the most extreme abuse of human rights yet devised – took place. It was seen to happen and yet elicited no effective riposte from the world outside. The explanation for this was given quite clearly at the time by Kofi Annan, the UN Under-Secretary-General for Peacekeeping. He explained that it was the post-Somalia syndrome. On 17 May 1994, Security Council Resolution 918 actually authorised the extension of the mandate of the UN peace-keeping force (UNAMIR) in Rwanda. But, following the loss of 18 US Army Rangers in Somalia, the USA was unwilling to have the UN launch an enforcement action to stop the genocide. And without US backing, as Morris has pointed out, the UN simply did not have the institutional capacity to respond. In order to avoid any form of responsibility for intervention, the word 'genocide' was never used during that period in any of the official texts or resolutions. The international community preferred to use the phrase 'humanitarian crisis' which implicitly places the logistics of aid well before the issue of protection.

In place of UN intervention, late and by no means replacing an early international response, the French government sent its own force, called Operation Turquoise. This venture was much criticised. And for good reason. It allowed numerous criminals to find a previously unimagined form of exile in the French-established security zone. Yet it did save lives within the safety zone that it created. It diminished the flow of refugees to Zaire, it protected some of those Tutsis who remained alive and some Hutus against revenge killings.

But the main thrust of the international response to the catastrophe in Rwanda was in refugee relief in Goma. There, once again, non-governmental organisations (NGOs) and international organisations set up a vast, sometimes even macabre, humanitarian bazaar. By spring 1996 the relief operation for the Rwandan refugees had cost at least one billion dollars – and to what end? Keeping people alive in a state of limbo and uncertainty which held no real hope for their future under the authority of those who represented the former regime responsible for the genocide.

Long gone, it seems now, are the heady days after the fall of the Wall when politicians spoke of a new world order. As Boutros Boutros Ghali put it, in his downsized Supplement to an Agenda for Peace, 'Collectively Member States encourage the Secretary-General to play an active role in this field; individually they are often reluctant that he should do so when they are party to the conflict.'

The Secretary-General also pointed out that 'It would be folly to attempt [an enforcement capacity] at a time when the Organisation is resource starved and hard pressed to handle the less demanding peace-making and peace-keeping responsibilities entrusted to it.' Indeed.

By early 1996, it was becoming increasingly clear that a catastrophe similar to that of Rwanda was about to befall Burundi. There was what some called a 'slow genocide' with dozens, sometimes scores or even hundreds, of people murdered every week. By April the numbers of deaths were estimated to have risen to 100 a day and, in all, at least 50,000 Burundians are thought to have been killed in the last two years.

Almost all these victims were civilians. The killings were often conducted in the most brutal fashion. About half a million had fled, abroad or internally. In March 1996 alone, 100,000 Burundians were displaced by violence – making a total of at least 600,000, or 10 per cent of the population. The civilian government hardly functioned any more. All effective power resided in the military. The Secretary-General called for one member of the United Nations to take up the role of lead country to make contingency plans for a humanitarian intervention should what I suppose one must call a 'swift genocide' suddenly gather pace in Burundi. He was not alone in his concern. By May 1996, US officials were pressing the UN to set up an intervention force for Burundi. But neither the US nor any other nation was prepared to assume the role of lead nation or to send any of their own troops.

At the UN itself, officials were frank about the quandary faced by the organisation. One senior UN adviser said, 'We are absolutely lost on Burundi.' UN officials argued, in the spirit of the revised Agenda for

Relief and vulnerability

MSF logistician Stephen Kamau Gatuma of Kenya was captured and taken hostage for more than a month. His story is a clear example of the vulnerability of civilians and of humanitarian staff in conflict situations.

On Saturday afternoon, 16 September 1995, the village of Panthou was attacked by the forces of Commander Kerubino Kwanyin Bol. He invaded the village and looted the 30 metric tonnes of WFP [World Food Programme] relief food that were stocked there. All humanitarian field staff were forced to flee under heavy gunfire. Stephen hurt his ankle, fell and crawled until he was found by Kerubino's soldiers. He was told that he could not leave since they feared he might be shot by SPLA [Sudan People's Liberation Army] soldiers and that Commander Kerubino would afterwards be blamed for killing an expatriate. The soldiers asked him where the radios were, but they were nowhere to be found, and had probably been taken by SPLA forces.

The next morning soldiers forced more than 50 villagers to accompany them as porters. Stephen witnessed depressing scenes of women and children carrying heavy loads of looted groundnuts and valuables. The villagers' cows and goats were also taken. Two young mothers, their breasts still dripping with milk, were taken without their babies. Only two men were chained and taken away as prisoners. During the move the soldiers were firing regularly so as to scare the neighbourhood and to chase people. In the different villages they encountered along the way, the soldiers forced the people to pull crops like maize plants and sorghum.

After three days of walking they arrived in the governmental garrison town of Gogrial. Stephen was told not to fear for his life and that he would be released in a couple of days. He was imprisoned in a small compound with a high fence so he could not see what happened outside. He received food twice a day, cigarettes and a small radio. During the following weeks he heard planes landing and taking off regularly.

On 17 October Kerubino visited his prisoner for the first time. He apologised for keeping him so long and said he would not ask for a ransom. He said that WFP food was not used as relief but was used to support SPLA soldiers.

On 23 October a plane brought Stephen to Khartoum, where he was released after interrogation.

While negotiating Stephen's release, both the SPLA and the Khartoum government denied that they had any links or contacts with Kerubino. The government claimed that he was an independent SPLA commander. In reality, Kerubino destabilises northern Bahr el Ghazal with the support of and on behalf of the Sudanese government.

September 1995, Bahr el Ghazal, Sudan

Peace, that the UN did peace-keeping operations – it did not provide rapid reaction forces to fight. One senior UN official said, off the record, 'The US is putting us under huge pressure in Burundi. The area is the Balkans of Africa. We all know it will blow. It could be any time. Members of the Security Council want to have clean hands.'

The UN also. No one wants to be blamed. So there is a sort of shadow game. But Burundi is so volatile that a Chapter Six operation would be impossible; it would have to be Chapter Seven. The Burundi army is well trained and the government has not given consent for UN intervention. Indeed, leading Hutus and Tutsis alike in Burundi had said that any UN intervention would be resisted by force. No wonder that UN Member States were slow to jump to a new mission. One further problem was to decide what would be the political objectives of intervention. UN officials were worried lest the Security Council pushed the organisation into another Yugoslavia.

But the key problem was where to find the troops. Some African governments said they would follow a lead nation, but whom? The USA, UK, France, Holland, Denmark, Russia could all do it, but none of them wished to do so. They all said they had 'no vital interests' there. No government was prepared to sell the idea at home. The UN would therefore have to find troops from countries which had other reasons for sending them – like money, or fear of refugees.

One possibility being explored was that Egypt might be the best hope for lead nation, with other African troops and lift capacity supplied by the USA and western Europe. It would not be a UN mission, but a multi-national force (as in the Gulf) with Rules of Engagement formed under Chapter Seven.

It was clear, however, that the most effective lead nation would be the USA. No one else had such planning or logistic capabilities, no one else would have a better chance of enlisting the participation of other governments. The US contribution could be restricted to leadership and logistical support. Even so, there was scant support for even that in Washington. There was also the real question as to whether the UN itself was capable of mounting any such operation, with or without the USA leading it. By 1996 the UN had the military capacity to analyse and prepare an operation, which it did not have in 1991. But there were fears that the organisation's financial crisis would destroy it again.

Moreover, there is still no proper system for the transfer of authority over battalions from national command to UN command. So when the bullets start to fly, nations begin to interfere in the chain of command. Thus governments told the UN when it could and could not move their troops around in Bosnia. An additional difficulty which is a consequence of the former is that New York is not an operational headquarters; it is a planning staff. It has no logistics to support a combat operation, and hires contractors to

Plate 1.4
Civilians are often targeted in revenge killings by rebel soldiers in Sierra Leone. Soldiers gather around a woman reportedly killed in a rebel attack in Bo, Sierra Leone. Photograph: Kadir van Lohuizen

deliver supplies. That works in situations like Cyprus, not in war zones like Bosnia. It means in effect that the UN can only do Chapter Six operations – peace-keeping – not Chapter Seven – peace-enforcing. Enforcement operations therefore have to be carried out by a multinational force – as in the Gulf, Haiti and NATO's eventual seizure of the military reins in Bosnia in August 1995.

Plate 1.5
A wounded man, a victim of Sierra Leone's civil war, is lifted into a jeep in the eastern town of Bo. Photograph: Kadir van Lohuizen

And here, contemplating the increasing reluctance of member nations to send their troops into situations where they might be placed at risk, more controversial solutions have been contemplated. When the French government announced Operation Turquoise in Rwanda, the venture was greeted with widespread hostility. The French were basically subcontracting a UN responsibility: in May 1994 the Security Council had authorised the UN force in Rwanda to protect civilians, but then had lacked the political will to find troops to carry out the resolution. The danger of subcontracting is, as Morris points out, that the UN may be turned into a flag of convenience. In the event the French adventure was more benign than its critics originally feared. And it is likely that subcontracting will become the norm.

Paradoxically, it was the UN Blue Helmets in Somalia who made the unexpected proposal of recruiting mercenaries in order to assume the mission of protection of UN and other relief activities against the omnipresent violence. But it was Sierra Leone, which offered the opportunity of observing such an operation in action. In this country, the small force of South African mercenaries called Executive Outcomes managed to patrol and control substantial sections of the country at minimum cost. Their presence and strength have acted as a deterrent against the rebels, limiting their harassment of the population. But one needs to note the circumstances behind the operation in order to understand why Executive Outcomes were hired in the first place. Paid for by the Freetown government, their essential role was to enable diamonds to be mined. Without such a commercial incentive they would not be there. And without such a revenue stream the government could not have afforded to employ them. If mercenaries are not to be deployed in the interests of diamond mines, then what? One Médecins Sans Frontières official suggested, ironically, that perhaps Coca-Cola would be a suitable employer of mercenaries – to open and protect new markets for the ubiquitous beverage. And after Coca-Cola, what about la Cosa Nostra? If armed protection is privatised and imported from foreign countries, it will only be carried out where there is profit potential. That is not an adequate answer to the questions posed by abuse of civilians around the world. What profit is there for anyone in sending a private army to Afghanistan? No commercial gain, just as there is no political gain. So it will not happen.

By early summer 1996, the International Committee of the Red Cross (ICRC) was at its wits' end as to how to protect civilians in countries like Liberia. Delegates wistfully recalled classic situations like Angola in which both sides allowed its delegations to work. Jean Daniel Tauxe of ICRC argues that the rest of the world must do more than just fund humanitarian operations. The Security Council must attempt to devise political solutions – no matter how controversial and difficult they may be. But again we get back to the paradox that while western television audiences want to stop seeing dying children on television, they don't want any of their own soldiers to die trying to help them by restoring order. That is a fundamental inconsistency (or even hypocrisy) by which relief agencies and governments are often hampered. Until there is some sort of resolution, the inadequate measures that we saw in Bosnia and in Somalia will continue to be the norm.

One controversial proposal that has been considered is a revival of the idea of Trusteeship. The American writer William Pfaff has argued, in *Foreign Affairs* and the *International Herald Tribune*, that a new colonialism is required. He suggests it should be taken on by the European Union, which, after all, includes all the old colonial powers. Others have suggested a revival and widening of the League of Nations mandate system for establishing responsible governments.

The problem, beyond the fact that it is hard to envisage Trusteeships on a long-term basis, is that it is highly improbable that the local powers will

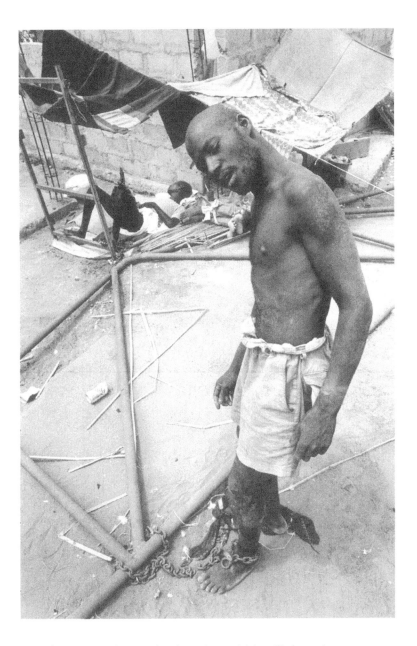

serenely accept an international tutelage which will deny them open access to all resources and complete legitimacy. This implies that one must be prepared to fight and kill for the benefit of those one wishes to help. It was with this in mind that UN forces intervened in Somalia following the Security Council's decision to adopt Resolution 814 in March 1993, its

Plate 1.6   The Institute for Traumatized Soldiers, Rangoon, Burma. Photograph: Jan Banning

Plate 1.7   Victims of Angola's civil war stand in the remains of a battle-ravaged building near the capital, Luanda. Photograph: Kadir van Lohuizen

most ambitious such operation to date. It was not just an attempt to assist a humanitarian operation or to assure peace. It was an attempt to re-establish the country's institutions by relying on a military contingent of more than 30,000 men and a budget close to one billion dollars. All that one can say is that the results obtained were hardly in proportion to the investment.

Cambodia underwent a kind of Trusteeship under the UN Transitional Administration of 1992–93 with the agreement of four of the country's political factions as a diplomatic means of resolving the national crisis. As a result, the country was better run than at any time in its history. There was a free press, the government signed the UN Covenants on human rights, and human rights groups began to flourish throughout the country.

Since the UN departure, however, there has been serious backsliding and many of the rights that the UN won have since been eroded, if not lost. Not all – there are still visible benefits from the UN's fleeting presence, but the experiment suggests that the UN can only skim the surface of ponds, rather than drag out the monsters that lurk below. As this goes to press, alarm is mounting in the international community that Burundi is sliding rapidly towards all-out disaster. There is talk, in the UN and in Washington, of preparing an international military force for humanitarian intervention, but nobody is prepared to volunteer troops, and for good reason: the objectives of such a force remain undefined. Both Hutu and Tutsi leaders have promised to resist any international intervention.

2

FRONTLINE MEDICINE
THE ROLE OF INTERNATIONAL MEDICAL GROUPS
IN EMERGENCY RELIEF

Mike Toole

•

DISASTER AND SURVIVAL: CONFRONTING THE PROBLEM

In the past two decades, wars in Afghanistan, Sri Lanka, El Salvador, Angola, the Lebanon, Somalia, Kurdistan and Rwanda, famines in Ethiopia and China, the floods of Bangladesh and Nepal, and earthquakes in Armenia and Mexico have all placed unprecendented numbers of civilians in need of disaster relief. An estimated 42 million people world-wide were dependent on humanitarian assistance in 1995, a dramatic increase of 60 per cent in ten years.

When the catastrophe is due to natural causes, the humanitarian response is straightforward: rescue operations, triage, medical and surgical treatment, rehabilitation and reconstruction. The initial goal is to reduce the suffering caused by injury, illness and malnutrition – and to prevent further deaths, all in a manner that helps restore the dignity and self-esteem of a community. In the case of armed conflict, however, survival means primarily refuge and protection. Victims need access to adequate food, shelter, water and essential services, such as medical care and sanitation. The normal provision of basic needs, such as the production of food by local farmers, the running of government or private health centres, or the supply of electricity by power companies is usually disrupted by the destruction caused by armed conflict and the associated mass population displacement.

Responsibility for the resolution of armed conflicts and the protection of civilians remains primarily with the United Nations (UN) and its member governments. Their record for taking action, however, has not been impressive, particularly in recent conflicts such as Afghanistan, Somalia and Chechnya. The direct provision of humanitarian assistance, an increasingly complicated and dangerous task, continues largely to be the role of non-governmental organisations (NGOs), although often coordinated with a UN agency. A striking feature of the international response to disasters has been the continued willingness of thousands of individual relief workers from around the world to travel vast distances in order to provide care to crisis-affected communities. These relief workers have largely been deployed by voluntary organisations such as Médecins Sans Frontières (MSF/Doctors Without Borders), Save the Children, CARE, Oxfam, Action Contre la Faim, Catholic Relief Services, Christian Aid, and the Red Cross.

Given the immediate goal of ensuring a community's survival, it is not surprising that health workers have been at the frontlines of relief operations. Most field relief teams comprise doctors, nurses, sanitary engineers, nutritionists, midwives and medical technicians, supported by logisticians and managers. To name just a few other than MSF, Médecins du Monde (MDM/Doctors of the World), the International Committee of the Red Cross (ICRC), International Medical Corps, Save the Children (SCF), Aide Médicale Internationale, Health Unlimited, German Emergency Doctors, the British medical relief agency MERLIN, and other similar medical groups have been able to mobilise thousands of these frontline relief workers from Rwanda to Bosnia, Tajikistan to Chechnya, Liberia to Bangladesh. Such often massive operations, however, are not without their drawbacks. It is therefore vital to examine critically the role of emergency medical relief organisations, their strengths and constraints, and the ethical dilemmas created for both agencies and individual relief workers by increasingly complex humanitarian crises.

EMERGENCY RELIEF – DEFINING THE ROLE

The world is small and full of kind people.
                         Zlatko Dizdarevic, *Sarajevo: A War Journal.*

The first response to a disaster is usually local. In 1991, when hundreds of thousands of Kurds fled the turmoil in northern Iraq and sought refuge in the mountains on the Turkish border, Kurdish communities in Turkey provided the first food supplies to the refugees. The second wave of response may or may not be initiated by the government of the country of asylum, depending on its attitude to the refugees and the adequacy of resources. Iran helped Kurdish refugees who fled there. The involvement of UN agencies, however, is often delayed. Organisations such as the United Nations Children's Fund (UNICEF), the United Nations High Commissioner for Refugees (UNHCR) or the World Food Programme (WFP) may have to depend on an invitation by the host government to intervene, or on politically driven decisions by wealthy donor governments.

The ICRC may be in a position to respond promptly because its mandate so clearly defines its responsibilities in conflict settings. Nevertheless, the ICRC's role depends on the ability to negotiate with the various belligerents, a process that may prove both lengthy and fruitless.

The NGO response, however, is driven by other factors. It can often be launched with minimal delay. The initial response at an MSF headquarters office, whether in Paris, Brussels or Amsterdam, is based on perceived humanitarian need. The first stage is exploratory: medical, logistical and coordinating specialists, both in the main offices and in the field, attempt to assess the scale of a disaster and identify the critical needs of the affected population.

Initially, the fact that foreigners – whether international relief agencies, diplomats, or journalists – recognise their suffering is often greatly reassuring and helps communities retain a sense of dignity and security. In a conflict situation, the simple presence of outsiders may discourage further human rights violations, whether massacres of civilians, rape, or torture. Or so it is hoped, often in vain. On occasion, the presence of expatriate front-line medical personnel has certainly helped restrain both government and guerrilla forces in conflicts such as Afghanistan or Sri Lanka. At other times, these belligerents have deliberately attacked civilians in the known presence of foreign aid workers, and even made attempts to kill or capture unwanted outside witnesses. During the Soviet–Afghan war, Red Army forces went out of their way to bomb clandestine hospitals run by international aid organisations, while Islamic militants mainly from the Arab world, fighting with the mujahideen (Afghan guerrillas), shot at ICRC vehicles, MSF volunteers and others seeking to provide humanitarian medical relief. In Angola, government and rebel soldiers both went out of their way to hit operations run by international medical groups seeking to work among civilians, regardless of location or political affiliation. The presence of relief personnel, UN peace-keeping forces coupled with global media coverage did little to thwart the fate of the Bosnian Muslims of Srebrenica at the hands of the Bosnian Serb military in 1995.

The presence of relief workers gives a sense of solidarity most valuable among populations where the world has either failed to recognise the gravity of their situation or has chosen for geopolitical reasons not to respond adequately. They often give eyewitness information to the world about the people's plight, and even advocate a more vigorous response by the international community. Such endeavours, however, have mixed results.

Early in the Kurdish crisis, relief agencies were loud in their advocacy of a safe haven for the refugees, with some success. In contrast, medical relief groups operating inside Iraqi and Iranian Kurdistan during the 1980s were able to draw only limited attention to the bombings and other atrocities levelled against civilians on either side of the frontier. Similarly, detailed information gathered by the aid agencies on high malnutrition and death rates among Somali civilians in early 1992 failed to mobilise the international community until much later that year, when the ill-fated military intervention was launched by the UN in conjunction with the United States military.

Talking is Better Medicine than a Tranquilliser

Leila is a displaced person from a small town in north-east Bosnia. Since February 1993, she has lived in Sarajevo with her two young sons. Her husband is missing. Living in Sarajevo means surviving in a small area of 5 by 15 km, and being a constant target for shells and snipers. She first visited an MSF counselling centre because of her children. Her 10-year-old son's unpredictable, aggressive behaviour was creating problems at school and her younger son, 8 years old, would not speak and frequently wet his bed. When asked about her experiences, Leila would say that she stopped thinking when she entered Sarajevo, that her past was gone, couldn't be changed and was not worth thinking or talking about.

Leila tried to avoid thinking about her experiences and suppressed the past, but the memories would often suddenly resurface. Her trouble with her past caused psychological problems for both children and Leila. By keeping herself occupied during the day, she could avoid thinking about her situation. But the nights were always an ordeal: she would sleep a few hours then lie awake for the rest of the night, her mind racing in a chaos of thoughts and worries. Sometimes she had nightmares in which she very realistically re-lived her flight. Questions about her husband were painful. When she thought about him she felt guilty and depressed.

Stories like this are often heard in the MSF counselling centres in Sarajevo. MSF started a psychological aid programme there in 1994 after establishing the need for it through extensive research. Some 10 per cent of those interviewed had a family member who needed professional psychological help.

Programme coordinator Kaz de Jong described how MSF helps people with post-traumatic stress syndrome: 'People confronted with traumatic events like this war need time to cope. They often cannot remember what happened to them. Because they have been in a life-threatening situation they have repressed their emotions. This is a normal reaction to an abnormal situation. Usually it takes a month or slightly more before the worst symptoms are over. If it lasts longer, we call it post-traumatic stress syndrome and then people need psychological help. In treatment session we let them relive the experience over and over again which helps them to accept the trauma. The key is: tell, tell and tell it again. Experience the emotions and share them with others.'

When Leila sought help at the counselling centre, she realised that her suffering was not unique. She learned to talk about her experiences and express her emotions. She and her sons attended 15 treatment sessions. Now the past is no longer a dark hole. Memories and experiences have stopped popping up at random. Although she regularly experiences grief and worries about the future, she no longer tries to block out the past. The children miss their father, but are allowed to express their worries. At school the older boy is being helped to socialise with other children. The younger boy has stopped wetting the bed, sleeps quietly during the night and has started to talk. The past remains a scar that often hurts. Leila's war experiences cannot be changed. But by allowing them to be a part of her life, she has found some peace for herself and her children.

Malou Nozeman, Press Officer, and Kaz de Jong,
Project Coordinator, MSF, Bosnia

Plate 2.1
As a leading medical
relief organisation,
Médecins Sans
Frontières provides
crucial vaccinations
in Benaco, Tanzania,
Photograph: Klaas
Fopma

THE DECISION TO INTERVENE

Decisions to establish full-scale relief operations depend on the perception
of need, the feasibility of providing assistance, and the availability of funds
(MSF maintains emergency funds for absolute necessities). Longer term operations require sustained funding. MSF has been relatively successful at this,
while avoiding over-reliance on single donors. Nevertheless, in certain situations such as during the early stages of the Somalia crisis, the operations
of MSF and other agencies such as ICRC and SCF did not have adequate
funding to meet the needs of the population.

When medical relief groups, both local and international, established
surgical services in Mogadishu during the civil war of 1991, an element of
hope was restored among the beleaguered and forgotten civilian population
of the city. The provision of basic health services helped promote the restoration of a sense of justice and dignity in a highly traumatised community
where men, women and children were at the mercy of ruthless warlords
and undisciplined fighting factions, who thought nothing of lobbing shells
into crowded markets or robbing and killing civilians for a small plastic
bag of sugar.

A primary goal during these early operations is the identification of the
most vulnerable groups in a population. Relief workers need to design

services to address the basic needs, such as focusing on the plight of thousands of unaccompanied children among the Rwandan refugees in Zaire. Notwithstanding the logic of focusing on the most vulnerable, this practice often means that many others in the community will not have their needs adequately met. This is not meant as a negative statement but merely reflects the dilemma inherent in a situation where scarce human and material resources need to be deployed in the most beneficial manner for the population as a whole.

EMERGENCY RELIEF: DETERMINING THE NEEDS

Relief requirements in an acute emergency setting are usually based on those conditions that most threaten survival, notably infectious diseases, nutritional deficiencies, and injuries. During the 25 years since MSF was founded, the agency's approach has broadened. It now seeks to address, for example, the root causes of higher mortality rates.

Some of the major developments within MSF during that time have been:

- the increased emphasis on the provision of the key determinants of health: shelter, water, sanitation and food;
- the use of sound information to identify and monitor the major health problems in a disaster-affected community;
- the attention given to disease prevention;
- the training of community health workers within the disaster-affected population.

These efforts to broaden MSF's relief activities from traditional medical relief to forms of assistance which might be better addressed by other international agencies such as Oxfam or CARE are not necessarily welcomed by everyone within the organisation. Some MSF medics and relief coordinators believe that the agency is moving dangerously away from its original goals by stretching itself too far, a view also shared by critical relief observers on the outside. But many others feel that the policy has led to a more effective response to emergencies and a more efficient use of resources. Nevertheless, the balance between resources allocated to the prevention and treatment of diseases is not always maintained. For example, in 1994 the initial reaction to the cholera epidemic in Goma, Zaire, was largely one of setting up treatment clinics and hospitals to care for the sick. Too little attention was given to improving the supply of clean water and adequate sanitation.

The needs and priorities of the affected community may prove considerably different from those identified by relief teams. For example, in 1985 some Ethiopian community leaders expressed concern about the focus of western relief workers on saving the lives of severely malnourished infants, rather than putting their efforts into treating the health problems of economically productive adults in the famine-affected population. Similarly, conflict-affected populations may consider themselves in urgent need of psychological support and the management of mental illness, but initially, at least, such services are normally not provided by relief agencies.

So relief organisations constantly have to reassess their methods according to the heterogeneous needs of affected communities and the evolving nature of disasters. Since 1994, various medical relief agencies have indeed established mental health and counselling services in Tanzania, Zaire and Bosnia, often based on information provided by population studies. For example, the decision to organise intensive feeding centres for malnourished children is usually based on a nutrition survey, whereby a sample of the population is weighed and measured and compared with an international reference. If the malnutrition rate is low, such centres will usually not be opened. This approach may prove sound in terms of public health planning, yet may exclude from the overall programme small but needy groups such as malnourished children in a large refugee camp. The use of epidemiological studies has contributed significantly to a better understanding of the impact of disasters on populations, more informed decisions on relief priorities, and, it is hoped, to the more efficient use of resources. However, the resulting selectivity of services may lead to a sense of relief aid being imposed by outsiders. As was recently witnessed in Somalia and Rwanda, relief agencies often tend to set up programmes based on their own agendas, including projects for the sake of projects or even for fund-raising purposes to take advantage of media coverage. Aid can thus take on invasive qualities. Assessments of disaster needs must include genuine and objective efforts to listen to the voices and opinions of those affected. While such approaches are clearly encouraged in current assessment guidelines, they are frequently neglected in the actual chaotic conditions of a disaster. In many cases, they tend to overemphasise quantifiable or measurable parameters. Political realities in the field, be they conditions imposed by donors on funding or decisions based on epidemiological or sociological studies, may all lead to situations where assistance programmes benefit only certain targeted groups in the population and exclude other needy groups.

NEW CHALLENGES

The relentless expansion of the global HIV/AIDS epidemic will increasingly complicate the medical, social and ethical context of emergency relief programmes. It has become clear, for example in certain central African countries where the prevalence of infection with the human immunodeficiency virus (HIV) is high, that many patients with infectious or nutritional diseases are not responding to standard treatment regimens. However, resources are usually lacking either to diagnose or to treat an underlying HIV infection, especially in an emergency setting. In the southern African country of Mozambique, for example, there are few reliably functioning diagnostic centres outside the capital. Moreover, the use of blood transfusions for surgical and medical patients has created the need for screening blood donors for HIV, often without facilities for providing counselling or maintaining confidentiality.

While MSF's experience in its formative years was gained largely in underdeveloped countries in Africa, Asia and Latin America, a profound change has occurred during recent years. Armed conflicts and subsequent humanitarian emergencies have increasingly taken place in cold regions such as Kurdistan, Afghanistan, Tajikistan and the Caucasus, and among industrialised societies in eastern Europe. This trend has posed new technical, logistical and cultural challenges. It means changes to the assessment by frontline medical groups of the needs of a population in crisis and the provision of assistance.

These changes are also related to the context of relief programmes, including troubling ethical constraints, the issue of inappropriate aid, and the nexus between humanitarian aid and long-term community development.

REALITIES AND MORAL CONSTRAINTS

We must pay and dirty ourselves with the meanness of human suffering. The dirty, repulsive university of pain.

(Albert Camus, *The Rebel*)

While the response to an emergency by MSF and other agencies might be simply described as one based on need, the reality is far more complex, notably in arenas of human conflict.

The establishment of a hospital to manage civilian casualties in a war zone such as Sri Lanka, Afghanistan or Liberia may mean dealing on a daily basis with combatants, many of them armed and often demanding priority care at the expense of civilian patients. Most medical teams insist, not always successfully, that combatants leave their weapons at the door when entering aid agency compounds. Establishing a relief programme in a conflict zone often places the agency in a position of moral compromise; to operate in that area may require payments to local militia for protection. In the case of Somalia, these militia were often simply groups of undisciplined armed bandits. In certain situations, too, relief workers seeking to maintain their services may have to turn a blind eye to blatant corruption, theft of relief supplies, and flagrant human rights abuses. The fighting factions during the war in Mozambique, government soldiers and officials in Sierra Leone, and both rebels and Russian troops in Chechnya all posed this moral dilemma. In the Rwandan refugee camps in Zaire, efforts to ensure the equitable distribution of relief supplies were thwarted by political and military leaders in the camps, many of whom had been

Adjaria, Georgia. Winter 1994

Every month we visit the hospital in Khulo district. The people in this area here are rough, extremely hospitable and they drink a lot – even according to Georgian standards. Khulo is the poorest and most remote district in Adjaria province. With snow on the roads it takes us at least four hours to get there from Batoumi.

Today we find that the hospital staff split up the drugs in two parts: one for the hospital and one for the director's room. We have experienced before that the director of Khulo hospital keeps the drugs that MSF distributes for himself, instead of giving them to the hospital pharmacist.

The part that he keeps in his room is supposed to go to smaller peripheral hospitals. But it is impossible to track down to whom he really hands out the drugs. Everything in the room is now stored on new shelves, even in alphabetical order. 'The drug mafia is getting organised,' I think cynically. I guess that our project here runs into the typical problems of any drug distribution programme. But in Khulo every problem seems magnified and I feel like Alice in Wonderland.

The people in this district are easy to get along with as far as eating and drinking are concerned, but at work they can be very distrustful and they are not very motivated. Despite all this we got the hospital staff to improve storage facilities and they organised a registration system. We have a good relationship with them that enables us to discuss medical topics such as problems over prescriptions.

But the impact of all our work is, to say the least, questionable. I am getting really confused thinking about cost-efficiency, reaching the target population, impact studies and everything else that I don't know enough about.

We go inside the cold hospital and upstairs to the surgical department. The chief surgeon is always happy to see us and claims to do a lot of good work with our drugs. But we have never seen more than a few patients at a time (maybe because it is too cold to be operated upon). We look at a fragile old lady who is still looking quite ill. She says that without MSF material she would not have had her operation. Then somebody takes my hand. It is the young peasant woman from the next bed. Our translator says that she wants to express her gratitude to MSF. She was operated on two days ago and did not have to pay for any of the surgical material. She will go home soon. I am pleased with her gesture, because I don't often meet patients who express their gratitude spontaneously; they are usually urged to do so by their doctor.

There are now only seven patients in the hospital, and it really is cold. I am happy to get back into our warm car. 'Maybe we should start distributing heaters,' I think to myself, 'but I am sure that the medical staff will just take them.' With limited time, staff, cars and money (in that order) maybe we should just stop working in Khulo. There are 30,000 people living in this area, but we seem to be unable to reach most of them. What are we trying to do here? This place needs more input than just a monthly visit. Maybe we should stop working in Khulo, or maybe we should work only in Khulo? My thoughts are running around in circles.

Our driver says that the best and probably cheapest way to help the population in Khulo is to repair the road.

Tineke Bronzwaar, Dutch medical doctor, MSF, Georgia

directly involved in the genocide in Rwanda. Consequently, an agency's decision to continue or discontinue assistance is often far more difficult than the original undertaking to initiate the operation. In the case of the Rwandan refugee camps in Zaire, MSF eventually decided to withdraw rather than continue to work in an environment where political leaders continued their violent oppression of the refugees and their exploitation of the relief programme to promote their own political agenda. The ramifications of this decision are still debated within the organisation.

For relief agencies, the ethical dilemma of whether to continue providing needed services to mostly innocent civilian populations in the face of ongoing human rights abuses or to withdraw in protest is often at the heart of today's response to complex humanitarian emergencies. Although the presence of expatriate relief personnel may be an important symbol of solidarity with a local population, there is also the danger that this presence may inadvertently provide legitimacy, power and wealth to a military or political faction controlling the area.

Plate 2.2 Frontline medical aid agencies provide key humanitarian relief, including emergency food supplies, to numerous war zones around the world. Local workers in Kenema, Sierra Leone, unload a transport aircraft. Photograph: Remco Bohle

INAPPROPRIATE AID

Useful relief assistance needs to be effective, efficient and delivered in a manner that helps restore the coping capacity of an affected population.

All three criteria can be breached by the provision of irrelevant and inappropriate aid. There are numerous documented cases of relief organisations, private companies, governments and individuals providing culturally unacceptable and nutritionally inadequate foods; inappropriate, expired, poorly packaged and even dangerous drugs; and other irrelevant relief supplies. While much progress has been made with the development and publication of guidelines for the donation of medical relief supplies, each emergency produces a new horror story.

In Goma, for example, one relief organisation chartered an aeroplane to deliver a huge shipment of a commercial soft drink, Gatorade, used by athletes, in the false belief that it could be used to treat persons with cholera. In fact, this product can prove dangerous if given to young infants. In addition, the product was not only bulky and difficult to store, but caused considerable waste and was not cost-effective when compared with the standard oral rehydration salts used to treat the diarrhoea and dehydration of cholera.

The World Health Organisation (WHO) field office in Zagreb, Croatia, has reported that of all the drugs received for use in medical relief programmes in 1994, 15 per cent were completely unusable and 30 per cent were not needed. By the end of 1995, 340 tons of out-of-date drugs were being stored and an incinerator had to be built to destroy them.

More attention needs to be given to controlling the problem of inappropriate relief donations. MSF, Oxfam and SCF are particularly active in this movement. In MSF's own field operations, reliance on unsolicited in-kind donations is avoided, and most supplies are obtained from the agency's extensive reserve stocks, at shipping points in Europe. Only supplies that are listed in MSF manuals and technical guidelines may be deployed for use in field programmes.

COORDINATION

A key obstacle to addressing this problem of inappropriate aid is the issue of authority. In the relative chaos of a relief programme, irresponsible or ignorant agencies may be able to deliver inappropriate relief supplies without reproach or restraint. Ideally, the government of a disaster-affected country should monitor the quality of assistance and ensure that inappropriate aid is rejected; however, either such governments lack the resources to do so or else governance in the country has totally collapsed. In the absence of government controls, the lead UN agency in place needs to exert this authority, but all too often does not.

During the Rwanda crisis, WHO was not always effective in asserting its role as an advisory institution with regard to international medical assistance. The result was that some of the NGOs involved in medical or nutritional aid had little idea how to deal effectively with cholera or other diseases brought about by poor sanitation, overcrowding and inadequate diets. Similarly, the international relief community is still waiting see whether the UN's Department for Humanitarian Affairs (DHA) will be able (or allowed by headquarters in New York) to assert itself in emergency crises, whether Afghanistan or Liberia, where more effective coordination among both international agencies and the NGOs could prove decisive.

In the long term, the problem can only be resolved by the relief agency community itself. This needs to be done through education of the public and the media and a willingness to maintain established standards of relief practice. An informed media could play a valuable role in monitoring and exposing agencies, companies and governments that persist in malpractice. There is no shortage of journalists with broad experience in covering wars and humanitarian crises and who report knowledgeably about such issues.

Yet very often, reporters and producers parachuted into these situations have little knowledge of the problems behind relief operations. They often fail to grasp the essentials behind the crisis or draw conclusions based on their own cultural backgrounds. These may have little to do with the situation at hand. At the same time, a good reporter, whether an expert or not, should be able to compile a relatively accurate assessment given sufficient time and access to sources. It is therefore in the interests of the medical relief agencies to ensure that journalists have the best possible access to those most qualified to discuss the crisis. This should include suggesting contacts among other aid organisations (including local NGOs) to ensure a broader and more accurate understanding of the issue and of what is best for the victims, rather than seeking to promote their own operations.

Ironically, the agency that flew the soft drink into Goma was praised for its ingenuity by a leading American newspaper, the *Wall Street Journal.*

PERCEPTION OF NEED

The need for ongoing assistance following the emergency phase may be difficult to evaluate objectively. The definition of populations in need is indeed a broad one. It could refer to the predicament affecting the majority of inhabitants of the economically underdeveloped nations of the world. This includes ethnic minorities, urban slum dwellers and the homeless, people displaced by development projects, such as dams, and environmental disasters such as deforestation and desertification, and dozens of other disadvantaged groups.

Most emergency relief agencies, including MSF, have tended to focus on populations perceived to be in acute danger. 'Acuteness', however, remains a relative term. Displaced populations in Mozambique, Angola and Sudan and refugees in Thailand, Pakistan and Ethiopia have received assistance for well over a decade, sometimes even twice as long. In such cases, where armed conflict, population movements and food scarcity have continued

Plate 2.3
Distribution of food is often chaotic: refugees in a Sierra Leone camp jostle to receive what little food is available. Photograph: Kadir van Lohuizen

over this extended period, it is relatively easy to justify the continued vulnerability of the displaced populations. On the other hand, those refugees safely protected in refugee camps or settlements in neighbouring countries (e.g. Vietnamese refugees in Southeast Asia) have in some cases continued to receive high levels of assistance while less visible, but possibly more needy, communities in other areas of the world do not.

The choice of aid recipients may be affected by geopolitical importance, media coverage, ease of access, and availability of donor funding. MSF must constantly resist the influence of these factors and return to its charter and its core values of neutrality and impartiality.

Any agency that has experienced rapid growth will be tempted to sustain that growth through opportunities presented by self-interested donors. However, a continuing influx of new volunteers whose experiences reflect the realities of the field can help ensure that these core values are sustained and protected.

DISASTER RELIEF AND REHABILITATION

While the impact of medical relief programmes on reducing death and illness rates has been well documented, these efforts have focused on the most extreme instances of populations deprived of their basic needs and their access to even minimal levels of health services. The fact is that such relief programmes scarcely address the overall problem of the poor health status of hundreds of millions of the world's inhabitants. Even in war-torn Afghanistan, there are many areas of relative peace which are denied basic health care because there is no emergency situation.

There is an overwhelming need to consider the restoration of public health services in a disaster as an integral part of restoring justice and dignity in a community. Such an idea assumes that health is a fundamental human right and that the international community has a responsibility to ensure that persons in all societies have access to the resources and services they require to maintain an adequate state of physical, mental and social well-being. This principle is not a new one. It has been endorsed by the World Health Assembly and by the landmark Conference on Primary Health Care in Alma Ata, Kazakhstan, in 1978.

In practice, however, it remains a far cry from most conventional approaches. It is vital to acknowledge that emergency relief programmes do little to contribute to the much-promoted concept of Universal Health for All. By definition, the objective of emergency relief, at least in conflict-related disasters, is to undo the harm caused by the world's diplomatic and political failures.

On a global level, much progress has indeed been made during the past few decades in reducing child and maternal mortality; however, this improvement has not occurred evenly throughout the developing world. In many sub-Saharan African countries, conditions have deteriorated significantly in terms of basic health indicators and access to health services. These are often the same countries as those affected by armed conflict and population displacement. At the same time, the wealthy countries of the world have actually decreased the amount of development assistance they provide, in particular to Africa, partly due to a shift in resources towards emergency relief, as the pool of overseas aid funds has remained stagnant.

The major global health programme funding agency is now the World Bank, whose approach may make a great deal of macroeconomic sense but in reality also may have contributed to the near collapse of public sector health services in many African countries. The increasing emphasis on user fees and the political and economic disintegration in countries like Zaire, Nigeria and Sudan have meant that many millions of the very poor have simply been excluded from basic health services. How, then, can medical relief-oriented agencies respond to these realities? It would be foolish to suggest that MSF and other similar specialist agencies abandon emergency relief in favour of development assistance. They have unique expertise in emergency work, for which there is a clear demand.

One practical response is to integrate health development activities more effectively within relief programmes and to take a longer term view of the needs of emergency-affected populations. Ensuring that community health workers are identified, trained, and given genuine responsibilities during a medical relief program provides lasting and useful skills for the community. In fact, this is increasingly a routine practice in MSF and other relief agencies.

An additional approach is to rehabilitate and restore the health infrastructure destroyed in armed conflicts. Several MSF branches have developed specific programmes to address the post-emergency needs of populations in countries such as Mozambique, Cambodia, Ethiopia, Georgia and Afghanistan. The maintenance of such programmes may involve forging relationships with local, indigenous NGOs, a process that requires a long-term commitment to institution building. Restored health services and locally trained health workers usually need ongoing support, supervision and resources to work effectively. Many agencies face a difficult challenge in finding the means to maintain that support beyond the crisis phase. This comes down to finding funds, which in turn relies on educating the public and government funding bodies so that they do not abandon emergency-affected communities once the media spotlight has dimmed.

CHRONIC EMERGENCIES

One of the most difficult situations in which to continue to provide assistance to populations in danger is that of the chronic emergency, particularly in countries such as Somalia, Liberia and Afghanistan, where governance has collapsed and where the international community has exhausted its generosity and interest.

In Angola, for instance, short-term, Band Aid-type assistance has been provided sporadically for more than ten years, but longer term assistance has been almost impossible for lack of donor funding.

In such settings, relief aid is often heavily targeted on particular areas affected by the conflict, while national development activities stagnate. Prolonged external assistance of an almost exclusively relief nature may result in over-dependency. It may also contribute towards draining communities of their energy and self-esteem as well as depriving them of opportunities to develop self-sufficiency and restore a sense of dignity. For international relief agencies wishing to involve local partners, identifying proficient groups, whether governmental or non-governmental, represents an enormous challenge.

THE RELIEF WORKER: WHY DO IT?

The fact that most of the major medical agencies successfully continue to recruit large numbers of highly motivated workers for demanding and sometimes dangerous humanitarian relief operations is morally reassuring. It implies that there is a widespread belief in the notion that the health of individuals and communities is a basic human right and that its restoration

during crisis situations is an integral part of establishing justice in a ravaged population. In reality, such apparent idealism probably grossly oversimplifies the motivating influences on a relief worker. The craving for new experiences and foreign adventures, dissatisfaction with one's current life situation, the need to make a difference, and the simple desire for personal change are undoubtedly important factors in the decision by an individual to join a medical relief organisation.

Plate 2.4
A child is weighed by a Médecins Sans Frontières volunteer in Kenema, Sierra Leone. Photograph: Kadir van Lohuizen

It is tempting to explain this motivation on the basis of the traditional medical ethic of providing assistance to all those in need. However, this ethical principle on the whole has been applied to the health needs of individuals in the immediate environment of the medical practitioner. Earlier this century, Dr Albert Schweitzer was one of the most articulate spokesmen for the idea of extending this medical ethic to include those of remote and different cultures. However, he used arguments based on a belief in European superiority that would find little sympathy in today's world. Many other individuals have been motivated by their personal religious beliefs to apply their professional skills to communities in distant lands.

The founders of MSF were a small group of French doctors who established a hospital for the victims of the Nigerian civil war in 1969. They acted at a time when the youth of affluent western societies was vigorously questioning the basically selfish and exclusive motives of the capitalist system. It was also a time when public respect for the medical professional was diminishing and the exigencies of busy medical practice had tended to push the medical ethic into a neglected corner of the physician's consciousness.

In fact, when these doctors founded MSF in 1971 they openly articulated the hope that the organisation could play a role in reversing the decline in the status of the medical practitioner. Consequently, an agency was created that attempted to institutionalise the idea that the medical ethic should contain no geographical, political, cultural, ethnic or religious boundaries. MSF's ongoing role in humanitarian assistance relies totally on the continuing availability of individuals who share those beliefs and values, even if other factors contribute to their motivation.

While the history of MSF has been dominated by medical professionals and has focused on health-related issues, the factors motivating its medical workers are not unique. Other relief organisations, such as the US-based International Rescue Committee (IRC), were created in response to the chaos and dislocation in Europe following the Second World War and the Nazi Holocaust. Oxfam was formed in response to the major famine in Greece during the Second World War. Other NGOs, such as World Vision and Caritas, have grown out of the religious beliefs of their constituents. The common element, at least at the individual level, is a belief that one's professional responsibility extends beyond the traditional confines of one's own immediate community. This idea, in turn, suggests a sense of promoting equity in geographically and culturally remote situations where populations are placed in peril because of natural or man-made events beyond their control.

ETHICAL DILEMMAS

In the course of his or her normal practice, a doctor or nurse trained in the western medical tradition usually responds to the medical needs of an individual patient in a setting where the patient voluntarily seeks care and chooses either the medical facility or the practitioner. Such choices are usually absent when a health professional becomes a relief worker in a disaster setting. It is the relief organisation that makes the big decisions: namely where, when and how to respond to identified disaster situations. Moreover, epidemiological assessments determine priority activities: agency policies, guidelines and medical kits determine what diseases will be treated and how; and political realities often determine the beneficiaries of relief programmes. One could argue that few decisions remain in the domain of the individual relief worker. How, then, does the average volunteer cope with these limitations?

In most circumstances, the volunteer is working in a community with real health and social needs far greater than those of the community that he or she has left behind at home. Provided this worker has relevant skills and receives adequate orientation, training and field supervision for the task, then he or she should make a significant contribution to the well-being of that community. When a community has such demonstrable needs as the Rwandan refugees stricken by epidemic cholera and dysentery in Goma, relief workers do make dozens of critical decisions daily and the contribution of each individual may be enormous. To rehydrate and save the life of a patient ill with cholera is tremendously rewarding for both the patient and the worker. When the scale consists of many thousands of affected persons, these medical activities may significantly promote the restoration of an entire community's well-being. Unfortunately, this is not always apparent to a global public that depends on the mass media for brief glimpses of populations in crisis and inadequate explanations of the response. For example, the world witnessed dramatic television images of Rwandan refugees in Goma dying of cholera and attended by a few overwhelmed and under-resourced relief workers. While the death toll within the first two weeks was indeed terrible, most of the media representatives left within several days of the initial crisis. They did not stay around to report on the dramatic success of the relief operation in the third week, during which daily death rates among the refugees decreased from about 6 per 1,000 to less than 0.5 per 1,000. This success was due to a well-coordinated focus on managing diarrhoeal disease, an achievement that went largely unreported by the media and unrecognised by the international public.

Nevertheless, serious ethical dilemmas constantly confront a health practitioner in a disaster setting. When epidemiological studies indicate that a population's priority health problem is one particular disease, such as cholera, this may mean that the immediate health needs of other individuals within that population are neglected by necessity. As was clearly evident during the Goma cholera epidemic, it was exceptionally difficult, if not impossible, for medical relief workers to provide individuals suffering from life-threatening tuberculosis with adequate attention. Although this makes perfect public health sense, it may create difficulties for a medical professional trained in the basic ethic of attempting to provide for the needs of every individual seeking care. The decision not to treat a patient with a curable condition – be it tuberculosis, mental illness, diabetes, or malnutrition – and the inability to diagnose and manage an underlying condition such as HIV infection, is often a painful one. The basic practical concepts that underlie disaster medicine triage – focusing care on those who can be saved, and doing the best for a population as a whole rather than for each individual – do not always fit neatly with the ethical concepts taught in most medical and nursing schools. Therefore it is imperative that medical relief agencies provide an alternative ethical framework that focuses on those actions that provide the greatest benefits for the majority of the population.

PRIMUM NON NOCERE

While the inability to address a broad range of medical conditions may cause frustration to many emergency medical workers, the discussion needs to be balanced by looking at another basic medical ethic: above all, do no harm. As noted earlier, the problems arising from the provision of inappropriate aid by agencies are a major issue. The dictum of do no harm, however, also applies to the individual health worker. Such harm could inadvertently result if a doctor, nurse, or nutritionist were to attempt to address each of the health needs of all the individuals in an emergency-affected population. How?

First, in the context of a large refugee or displaced persons camp, relief workers would find themselves perpetually locked into outpatient clinics and hospital wards, spending all their time on treating diseases in a relatively small proportion of the overall population. In so doing, they might possibly neglect the more effective preventive activities such as immunisation against measles, thereby creating a situation where thousands of children remained vulnerable to a life-threatening, but eminently preventable, disease. Their efforts to cure the conditions of relatively few individuals might lead to epidemics and large numbers of deaths among a majority of the population. Consequently, they might do more harm than good for the majority.

Second, attempts to treat a broader range of medical conditions would require enormously complex and expensive logistical support to provide the necessary range of drugs, medical and diagnostic supplies. By focusing on those conditions that cause most illness and deaths in a community, drugs and other medical supplies can be efficiently provided in pre-packed kits.

Third, commencing treatment of a chronic illness such as tuberculosis in an unstable and uncertain setting may actually do severe harm to the community. This particular disease requires at least six months of daily treatment with three different drugs. If treatment is discontinued because of lack of supervision, sudden departure of the patient to another region, or disruption to the supply of medications, then resistance by the TB organism to the drugs may develop and the ability to treat future patients in the community may be jeopardised.

Finally, if health workers persisted in attempting to offer the kind of individual attention that their sense of medical ethics sometimes demands, they would have little time to go out of their hospitals and clinics into the communities, to monitor and assess the problems of those communities, to identify and train local health workers to carry out basic primary health-care services, or to evaluate the impact of their efforts. They would soon be out of touch with the real situation in the community.

WORKING IN A CONFLICT ZONE

While the daily dilemmas created by the tension between caring for sick individuals and ensuring a healthy population may be frustrating, they become minor irritants when a relief worker is confronted by the demands of working in an area of armed conflict or massive human rights abuses. This is a challenge faced both by the agency and by its individual volunteers.

In Bosnia, relief workers witnessed appalling examples of torture, sexual violence, murder, slave labour, ethnic cleansing and the confiscation of medical supplies. In Somali refugee camps in Kenya, enormous numbers of women were raped. In Rwanda, medical workers witnessed their patients being murdered in their hospital beds. In the camps of eastern Zaire, refugees were murdered on a daily basis merely because they expressed a desire to return to their homes in Rwanda. In Mozambique and Liberia, boy soldiers have been forced to murder their own families.

Such atrocities and systematic human rights violations against civilian populations occur more and more frequently alongside humanitarian relief operations. In such settings, humanitarian assistance is incompatible with maintaining silence. Both relief workers and their agencies should be made more aware of the need to highlight human rights abuses. They can help to make such information available to specialist human rights groups (particularly if there is a danger that on-the-ground monitoring might jeopardise the role of the individual relief worker) or squarely face their responsibilities as advocates on the international stage, a challenge now embraced increasingly by a handful of emergency relief agencies.

FRONTLINE MEDICAL RELIEF:
NEW DIRECTIONS, NEW DILEMMAS

Much progress has been made during the past 20 years in the provision of appropriate and effective humanitarian assistance to populations in crisis. There has been significant improvement in the accurate assessment of public health needs, the collection of ongoing information on the health and nutritional status of populations, the focus on disease prevention and community health, the evaluation of the impact of relief programmes, and the training of both local and expatriate relief workers.

Thousands of individuals from an ever-growing number of countries continue to make the personal decision to engage in assisting remote communities in acute distress. This commitment is in itself a dramatic illustration of the ability of many medical professionals to extend their ethical responsibilities from their own immediate circle to distant and unfamiliar cultural groups. That this commitment also creates new and complex ethical dilemmas in their fieldwork is part of the reality of relief assistance in areas of ethnic, religious and political conflict.

But the ethical problems related to working in a conflict setting are not so easily addressed, either by an agency or an individual. While excellent guidelines on a range of technical issues have been developed and disseminated, no such guidelines exist to address the ethical challenges of relief work. Such guidelines cannot be developed in a global context where a consistent set of moral principles does not exist to guide the international response. NGO programmes will continue to encounter these ethical compromises unless the international community can develop a more comprehensive and consistent approach to the prevention and resolution of armed conflict.

No international medical relief agency alone can adequately address the enormous needs of populations in settings where either a chronic conflict situation exists or where governance and public health services have totally collapsed. This is a massive problem that requires the long-term support and commitment of the international community, in particular the governments of wealthy industrialised nations. Nevertheless, NGOs can be effective advocates in the international arena; their experience and effectiveness in caring for populations in extreme danger has established their credibility. With a united and consistent voice, they can also be effective in mobilising international resources to ensure that health as a human right is more than a mere fantasy for the world's poor.

3

INTERNATIONAL HUMANITARIAN ACTION
GROWING DILEMMAS AND NEW PERSPECTIVES
FOR THE TWENTY-FIRST CEBTURY CENTURY

*Pierre Gassmann**

•

As groups and individuals involved in international humanitarian action on the ground, we are all confronted increasingly by intolerable dilemmas. The cost of our perseverance in seeking to help innocent civilians who suffer at the hands of their own people appears much too high.

We accept compromises in the hope of sharing our values in the long term. We are disappointed by the cynicism and inertia of the political leaders. We feel resentment as the international community produces ever more politically correct euphemisms in order to avoid calling war by its proper name. We are anguished by the blind violence that we are forced to confront. And we are saddened by the sacrifice of our colleagues who have lost their lives in the line of duty.

No new order has emerged to replace the apparent equilibrium produced by the Cold War both between East and West but also the Third World. We no longer seem confronted by so-called 'just' wars. Nor can we continue to attribute the current weaknesses of our doctrine of neutrality to the failure to act by governments or to the insufficiencies of the humanitarian community. This deterioration of conditions only obliges those among the international relief organisations who seek to remain neutral to embrace even more uncomfortable humanitarian policies. It is forcing them to adopt a far too scrupulous respect for the abusive demands of illegitimate sovereignty. From the humanitarian point of view, it obliges them to do far too little by restricting their efforts to remain present on all fronts and all sides. The satisfaction of being able to provide a 'just' sense of humanitarian action has given way to confusion in the face of too great a diversity of causes. We are caught up in an atmosphere of omnipresent barbarism, in which different belligerents make extravagant political promises.

---

*The views expressed in this essay are those of the author and not necessarily of the International Committee of the Red Cross. This piece was written shortly after three ICRC representatives were killed deliberately by unknown gunmen in Burundi in early June 1996.

International humanitarian action has never been easy, either to justify or to put into practice. However, it has always seemed possible to reinvent it. This has been one of the objectives of the Red Cross Movement and among religious charitable organisations, which have always sought to perform their duties whatever the political context.

Numerous critics have strongly condemned the form of humanitarianism practised by the ICRC and religious organisations, which seeks to alleviate wrongs while acting as an intermediary in the violence. It is considered misguided because it recognises war as an integral part of the human condition. Humanitarian action, these critics maintain, should make its priority the termination of strife. It needs to use all means at its disposal to search for political solutions and peace. This means combining humanitarianism with politics by becoming more directly involved in the implementation of so-called 'corridors of tranquillity' and 'spheres of peace' or the adoption of humanitarian cease-fires and 'Agendas for Peace'.

Having acquired this political albatross, however, international humanitarian action is now being restored by all sides, whether on the left or the right. Belligerents are now manipulating humanitarianism in their own interests by forcing relief agencies to abide by their own terms. And by doing so, they are succeeding in perpetuating the dependence of their captive civilian populations on their own political machinations; these are the very civilians whom these belligerents deliberately abuse or attack as a means of perpetuating their messages of terror and control.

Only until recently could one regard the impotence of humanitarian engagement as the result of a guilty silence. Such criticism was directed in particular against those relief agencies who sought to justify the need for public discretion, if not silence, with regard to certain aspects of their work. These agencies feared that by being overly vocal they might lose the right granted to them by the international community to help and protect certain categories of people. As a result, organisations which sought to undertake humanitarian action without necessarily providing systematic public testimony of abuses were condemned morally. According to their critics, it is the obligation of all those involved in humanitarian action to make political denunciations.

Nowadays, it is virtually impossible for human rights violations or excesses of violence to remain unpublicised. There are sufficient numbers of humanitarian organisations and human rights advocates willing to alert both the media and the general public to such occurrences. By their efforts, these groups are able to influence public opinion and thus persuade civil society to influence their governments. In fact, they have succeeded so well that the suffering of far-off peoples has now become a common fact of life.

Given such global attention, one can now question, quite justifiably, whether the international press still has any impact whatsoever on the attitudes of belligerents. All too many political leaders, warlords, faction chiefs, or militia commanders involved in conflicts or humanitarian crises are adept at manipulating the media for their own ends. At the same time, the publicity given to humanitarian action by the media has led western governments and the military to become directly involved in it too, using it as a means of projecting a positive image and avoiding any genuine political involvement in the search for political settlements. Large-scale humanitarian operations may appear to reflect a global willingness to act, but they resolve little other than provide short-term 'Band Aid' solutions.

Another acute problem is the growing dependence of traditional humanitarian organisations on the financial contributions and support of governments. Few so-called Non-Governmental Organisations, or NGOs, can truly claim to be non-governmental. This dependence prevents them from acting freely and from focusing solely on the objective needs of the victims. There is a growing awareness, however, that for humanitarian action to be effective it must remain totally independent of all political interference by the donor governments. The impetus for humanitarianism can only lie in the domain of public concern. What is needed now is for civil society to be mobilised as a counterweight to government manipulation of humanitarianism as an instrument of policy.

One of the consequences of this emerging school of thought is a growing solidarity among numerous non-profit-making organisations. The need to remain free of all political interference has become an urgent issue. Few humanitarian organisations with international responsibilities, however, shrink from accepting government funds. In fact, many cannot afford to cover the costs of their operations without such support. At the same time, donor governments now have a choice in their selection of humanitarian partners. They can always find one ready to serve as a front for projects inspired purely by political objectives. It is not surprising that humanitarianism is being taken over by governments today, in much the same manner as it was during the Cold War by various regimes and armed groups when relief organisations sought to protect and assist victims.

Many are now questioning, finally, whether traditional humanitarian organisations have gone too far in their reverence for political leaders or groups in control of countries or areas where the humanitarian dramas of the past few decades have taken place. Are these so-called leaders not using and abusing the masquerade of sovereignty as a means of asserting their positions in the international arena. Was it right for the international humanitarian community to impose their own philanthropic Judaeo-Christian values on peoples who had freed themselves from their colonial yokes? Shouldn't we have provided them instead with a network of private organisations so that they could create a civil society with its own values? Having contributed in this way to a weakening of the already fragile legitimacy of these states, humanitarian organisations are today complaining about the incapacity of these very states to protect their own citizens and to respect their international responsibilities. Furthermore, the attitudes in these countries can only increase the incomprehension, even hostility, that now exists towards our efforts at solidarity.

Under such circumstances, we are all tempted, almost daily, by renunciation and by the need for reward. During the most recent clamour in the media, opportunities were lost because of our incapacity to act. Or worse, they appealed to our desires for force to be used in order to re-establish order and to enable us to work. However, the logic of renunciation, a reflection of our own social problems back home, remains incompatible with our humanitarian motivation. This is valid not only for the organisations which have chosen to work in these situations but also for ourselves as individuals.

Some of us will accuse humanitarian organisations of an egotistical desire to perpetuate their bureaucratic existence at any cost. A substantial number of aid agencies have recently been accused by both the press and various public reports of indulging in relief operations more for fund-raising purposes than out of humanitarian concern. Others will see the unscrupulous exploitation of the international humanitarian community by certain armed forces to improve the quality and quantity of their equipment as a crippling threat to humanitarian action. Increasing numbers of armies are seeking to involve themselves in international humanitarian operations for PR purposes but also as a means of training. Whether soldiers can do the job better than humanitarians is another matter, and one which is being currently debated. Such criticism, however, needs to take into account the enormous progress that has been achieved along the long road that has brought us to where we are today.

Should one argue that the neutrality of humanitarian action today no longer obliges certain aid organisations to refrain from speaking out, but that all human rights violations, no matter where, should be denounced? Who can honestly claim that humanitarian action, even if well-meaning and not imposed by force, cannot be validly denounced by anyone as benefiting the enemy or as a violation of sovereignty?

Should we deplore the involvement of donor states in the humanitarian arena? It is up to us to ensure, through our political institutions, that whatever is done is according to principles and criteria compatible with humanitarian ideals and impartiality. It is up to us to convince governments that the humanitarian action proposed by civil society merits the support of states, but that we need to remain autonomous in its implementation. It is up to us to persuade donors to find the political solutions and to adopt the necessary measures to ensure international security by whatever means are at their disposal. With this in mind, it is interesting to note that in the face of political expropriation, particularly by the donor governments, the heads of the major United Nations organisations are increasingly closing ranks to criticise the decisions of the General Assembly and the Security Council. Lacking independence, they are at least seeking greater autonomy.

Should we not be encouraged to note that from now on violations of international humanitarian law and human rights will be taken into account, even if only selectively and timidly? That such abuses are seen as threats to international stability and security? And that they are serving as an argument for intervening in the internal affairs of states which have severely violated the rights of their own peoples?

Why shouldn't we welcome the fact that western armies, as part of a multilateral intervention force, are prepared to deal with humanitarian problems? They will learn soon enough of the limitations imposed by the chaotic realities of relief work. With time, however, the restraints of humanitarian reflection can only contribute towards a better understanding and improved application of the law limiting military intervention and the use of certain weapons.

It is also a positive development that the media hype over certain humanitarian interventions, coupled with the criticism which it engenders today, will lead to greater political responsibility among the donor countries. This will also enable them to distance themselves from purely reactive policies. It would allow a return to a more balanced search for medium- and long-term solutions.

Lastly, the increase in the number of humanitarian organisations, despite the confusion it creates, can be beneficial in more ways than one. The breadth of the problems, and also their diversity and complexity, demand more than just knee-jerk responses. No single international humanitarian organisation can pretend to respond to all the challenges and nuances of every situation. It is the existence of numerous options that allows some to follow the particular path they have chosen, others to implement the formal mandate they have received from the international community.

By the same token, some organisations can, and perhaps should, remain silent because they believe that the only way they can stay with the victims they seek to assist is by being discreet. But they also know that other organisations, whether human rights groups or frontline relief agencies, will appeal to the public conscience. This enables such groups to pull out of an operation if they feel that their involvement is being exploited to the detriment of the victims they are trying to help. Yet the more outspoken groups that pull out know that their gesture is only condemning those they seek to denounce: they are aware that someone else will continue to ensure the survival of victims they have been forced to leave behind. For all concerned, however, there is the hope that the impact of those who have departed may also make the position of those who have stayed more tolerable.

4

THE PLIGHT OF
THE WORLD'S REFUGEES
AT THE CROSSROADS OF PROTECTION

François Jean

•

In a world in a state of upheaval, the plight of refugees is a tragic illustration of the convulsions that plague the planet. They are evidence of the war, famine, and oppression that force millions of displaced people out on to the roads of exodus. Over the past few years, the multiplication of conflicts and violent situations has swollen the ranks of refugees and displaced populations to over 50 million people. The magnitude of this forced displacement of population and the growing numbers of asylum seekers constitutes a tremendous challenge to the international system of protection for refugees. Created at the end of the Second World War, this system is now at a crossroads, and the current evolution of refugee policies in Europe and the United States presents a crucial question: are western countries, which have a strong influence on international norms, willing to continue to ensure the protection of refugees?

THE SYSTEM OF PROTECTION FOR REFUGEES

The refugee question is by no means a new one, for human history is full of episodes of people forced to leave their homes. But it was not considered a specific social phenomenon until the end of the sixteenth century. The word 'refugee' was in fact coined in 1573 in regard to the Dutch Calvinists who fled persecution in their Spanish-dominated homeland and were taken in by their French brethren. Despite the fact that they were Protestant, they were protected by the king of France, then hostile to the king of Spain. The etymological history of the term reveals that, from the start, the refugee was identified not only by the persecution he or she suffered, but also by the sense of responsibility he or she evoked in others. Refugees have always existed, but their protection has always depended upon questions of specific solidarity and of political interests.

The twentieth century marked a change from *ad hoc* responses and selective solidarity to a universalisation and institutionalisation of the refugee problem. With the creation of the League of Nations at the close of the First World War came the notion of the international community's global responsibility to aid and protect refugees. After the Second World War, the creation of the United Nations High Commission for Refugees (UNHCR) and the signing of the 1951 Convention on the status of refugees introduced the international system of protection of refugees as we know it today. The Convention defined as a refugee any person who, owing to a well-founded fear of being persecuted for reasons of race, religion, nationality, membership of a particular social group or political opinion, is outside the country of his nationality and is unable or, owing to such fear, is unwilling to avail himself of the protection of that country. Moreover, the Convention insists on the essential principle of non-*refoulement* (driving back), whereby no refugee can be compelled against their will to return to their country of origin, where they might risk persecution.

Over forty years after its adoption and ratification by over 120 states, the 1951 Convention remains the cornerstone of the system of refugee protection, but it is marked by the context of its origins. The Convention's definition, founded on an individual approach concentrating upon persecution and discrimination, reflects the concerns of a Europe marked by the aftermath of the Nazi nightmare and already threatened by the Soviet system. From the beginning of the 1950s, Europe was locked into two opposing blocs, and in this Cold War context, refugee became synonymous with dissident; most refugees were fleeing totalitarian regimes to seek asylum in democratic countries. The solution then was to relocate them definitively in Europe or the USA and to confer upon them legal status and rights closely similar to those of the citizens of their host country. Asylum policy was all the more liberal since, in the prevailing climate of ideological confrontation, eastern European refugees were greeted with sympathy and were apt to blend easily into the host population because of common cultural affinities. In fact, until the end of the 1950s, the refugee problem was essentially an intra-European one of movement from east to west. Although it pretended universality, the 1951 Convention actually only applied to Europe, and it was not until the New York protocol of 1967 that the UNHCR mandate was extended to the rest of the world.

In the early 1960s, wars of national liberation and the first conflicts in the newly independent states of Asia and Africa began to provoke important movements of refugees. After decolonisation, the UNHCR, as well as the World Bank and other United Nations organisations, turned its attention to the Third World and had to adapt to a new situation of south–south movements of populations and large-scale exodus caused by war and insecurity. Unlike dissidents from behind the Iron Curtain who arrived individually at the portals of the West, refugees from the south are collectively fleeing situations of conflict and usually seeking temporary haven in a neighbouring country. Thus the UNHCR's area of competence was expanded by the UN General Assembly to encompass mass exodus of populations, and the definition of refugee was *de facto* enlarged from persecuted individuals to mere victims of violence. This expanded definition was more or less formalised, for Africa and Latin America, by the Convention of the Organisation of African Unity (OAU), in 1969, and by the Carthagena Declaration of 1984, both of which recognised as a refugee anyone fleeing from war, insecurity, or massive violations of human rights. The international community responded to these mass exoduses of the Third World mainly by providing humanitarian aid in the refugee camps.

FROM REFUGEE CAMPS TO LASTING SOLUTIONS

For the last three decades, the majority of populations fleeing war, famine and repression have been from countries in the south, and they seek refuge in neighbouring countries. At the end of the 1970s, the hardening of the East–West confrontation and the multiplication of low-intensity conflicts caused major refugee movements in Afghanistan, Southeast Asia, Central America, and the Horn of Africa. Since the end of the Cold War, the great concentrations of refugees are found around countries in conflict (Burma, Afghanistan, Tajikistan, Azerbaijan, Georgia, ex-Yugoslavia, Chechnya, Liberia, Somalia, Sudan, Burundi, Rwanda, etc.). These refugees are usually placed in camps organised in the host countries with the assistance of the international community.

Faced with such immense exoduses as those of the Kurds of Iraq, the Rohingyas of Burma, or the Rwandans, the international community must act quickly and effectively in order to meet the needs of helpless populations driven to the borders of countries in conflict. This assistance is all the more necessary since, although the refugee camps are conceived of as temporary, sometimes these chronically precarious structures, which are totally dependent upon international aid, are perpetuated for up to ten or fifteen years. Effective assistance in the form of rapid delivery of aid and prompt arrival of humanitarian teams in a crisis situation is also essential to the acceptance of refugees by a host country that might otherwise be tempted to repatriate them forcibly. The central role of the UNHCR in the coordination of international assistance is certainly a factor permitting it to assume the mandate of protection the international community has bestowed upon it.

The protecting role of the UNHCR is all the more important since the refugees are the victims of violence and war, and a simple passage over a border does not isolate them from the tensions and power struggles that afflict their homeland. And refugee camps are not wastelands where crowds of victims who know no past drift to a haven. They are complex societies where force often reigns and control is, more often than not, in the hands of politico-military movements. Over the years, certain refugee camps have become humanitarian sanctuaries and consequently a factor in the perpetuation of conflicts. Guerrillas find political legitimacy in the camps, through

their control over the refugee populations, and an economic base, since the camps are supplied with aid and a ready reservoir of fighters. In the 1980s, the sites controlled by the Contras in Honduras and the Khmer Rouge in Thailand or, more recently, the Rwandan refugee camps in Tanzania and Zaire, are perfect examples of the manipulation of refugee aid by politico-military movements.

The realisation of the limits and the perverse effects of the indefinite prolongation of a humanitarian status quo in refugee camps raise the crucial question in regard to possible solutions beyond immediate aid. In the past few years, thinking on this problem has evolved considerably. Traditionally, there were thought to be three lasting solutions to the refugee problem: integration in the host country, resettlement in a third country, or repatriation to the country of origin. Today the first two options – integration and reinstallation, in other words asylum in Western countries – are less and less considered as current solutions.

Even if many countries, particularly in Africa, remain relatively open to refugee movements, integration in a host country is regarded less and less as a realistic solution. In fact, in most cases, the host countries are poor, fragile, and have neither the means nor the cohesion required to integrate thousands of refugees. Their reticence is all the more evident since northern countries, which are supposed to set the example of respect of asylum rights

Plate 4.1
Internal refugees displaced by the civil war in Sierra Leone fleeing along the main road from the eastern town of Bo to Kenema. Photograph: Kadir van Lohuizen

and had for a long time made it possible for certain categories of refugees to seek refuge on their territory, are now increasingly reluctant to do so. In the last decade, perceptions have changed immeasurably in western countries: refugees who had a political significance during the Cold War (they voted with their feet) and a positive image (they chose liberty) are now considered undesirables by increasingly isolationist countries haunted by the spectre of mass immigration.

The evolution of the question of the Vietnamese boat people illustrates this change in perception. The Vietnamese boat people, recognised *a priori* as refugees during the first international conference on Indochinese refugees in 1979, were considered potential immigrants a decade later. The Global Plan of Action adopted by the international community in 1989 instituted a policy of so-called human deterrence to prevent people from leaving Vietnam and began to permit forced repatriation of those boat people who were not recognised as refugees.

Events in the Caribbean, in the context of Cuban–American relations, the last stronghold of Cold War logic, confirm this change of policy. Since the summer of 1994, the Cuban boat people have lost their political heft, their symbolic status and their media visibility; the American Coast Guard now systematically turns them back at sea, in flagrant violation of the principles of the 1951 Convention. The history of these two groups of people who long enjoyed a privileged status allowing them to seek asylum in Europe or the USA is indicative of the climate of rejection which continues to grow in the West. Cold War certainties have given way to a profound disquiet in the face of upheavals all over the globe and the fear of mass immigration. This reticence has increased as south–north movements of refugees towards western countries have been added to south–south flows which still continue to drain off the bulk of refugees. Although those seeking refuge in the north still represent only a marginal percentage of the populations of southern refugee camps, a perceptible increase in the numbers of those requesting asylum in Europe at the end of the 1980s has motivated a profound policy change confusing the refugee question with migration problems. Western countries try to discourage asylum-seekers from knocking at their doors and confine themselves to an increasingly restrictive interpretation of the 1951 Convention. The refugee question, once considered in the light of human rights, is now seen as a threat of immigration. The magnitude of refugee movements and the growing number of asylum-seekers in western countries has catalysed a profound change in refugee policy in the north as well as the south. The chronic state of camps reveals the inadequacy of aid policy in the south, and the reticence of host countries marks the limits of a policy of reinstallation in the north. The tandem of aid/resettlement which has been the cornerstone of refugee policy for three decades has now been replaced by the key words 'repatriation' and 'prevention'.

Letter from Zekira

The following is a letter from a refugee at the Kuplensko refugee camp in Bosnia to her family in Ljubljana.

My dearest grandfather and father,

This letter has been written by your daughter-in-law and your grandchildren. We are well and healthy, and hope you are the same. Only great fear and depression has forced us to write you this letter, which will be posted by a good friend of ours.

We were informed that you tried to visit us, and we know everything about the attitude of the Croatian police towards you. You were simply forbidden to enter the camp to visit us. Only our Lord knows if we'll ever have a chance to see each other again!

This might be the last letter you ever receive from us, so please put it somewhere safe to remember us by.

The living conditions here are equal to the conditions in the worst of camps. And recently the joint force of the Croatian police and the Croatian army made a blockade around our camp. They probably want to deport us back to where we came from. You know as well as I do that that means certain death for us. Anyway, we lost everything, all of our material property. The Fifth Corps of the BiH [Bosnian Government Forces] army burnt everything to the ground. Why do we have to return home to pass away when we can die here just the same?

The Croats forbid the UNHCR to supply us with food, and we have not received any sugar for more than three weeks. There are worse things than the shortage of food, but it is better not to write about them. All the Croats are well armed and they built numerous blockhouses around our camp. It's so cold and they forbid us to collect wood to make fires.

We don't sleep during the night for fear that the Croats will come and take our father away. Every single day, they take some of our men to Kladusa and we never hear from them again. They simply disappear, vanish. What if they kill our father now, what will become of us! We do not want to let our father go. If we have to die, we'll die here, out of hunger or out of indescribable fear.

My dearest grandfather, you remember we used to go to school; now we have forgotten nearly everything we had learned. There has been no school here for such a long time. Please, dear grandfather, help us come out of here. We are so afraid of all these armed men around us. You cannot imagine the sight of so many armed men on the blockhouses.

Please, dear grandfather, help us. If there is nothing you can do, pray for us to endure. To tell you the truth, I do not think we can endure much longer.

Zekira, Kuplensko Refugee Camp, Bosnia

Rohingya repatriation

In early 1992 some 260,000 Rohingyas fled Burma in the face of the government's increased militarisation, destruction of their villages and confiscation of their land, forced labour, intimidation, and degrading and inhumane treatment. They settled in 19 camps in Bangladesh. The government of Bangladesh worked with UNHCR to set up the camps, but intended from the start that they would only be temporary. By the autumn of 1992, the Bangladesh government started to force refugees to return to Burma. Camp authorities intimidated refugees and arrested their leaders. As a result UNHCR disengaged itself from the camp administrations.

Despite the fact that the situation in Burma did not change considerably by 1993, UNHCR stepped up its repatriation efforts. Following discussions with the government of Bangladesh in December 1993, UNHCR announced a mass repatriation plan based on individual interviewing. After the first sessions of interviews in Kutapalong camp showed only a 26 per cent willingness to return to Burma, UNHCR changed to a bolder approach. Information sessions became promotion sessions with the message that it was time for the refugees to go home. UNHCR argued that the situation was 'conducive' to their return.

In a draft agreement with UNHCR, the Bangladesh government argued that the Rohingya refugees should not be treated as refugees any more and that they should be sent back as soon as possible. The Bangladesh government also pointed out that the willingness of the refugees to return should be weighed against the Burmese government's sudden willingness to allow UN staff to assist the repatriation process and monitor the situation upon the return of the refugees.

The UNHCR then moved from private interviewing to a massive registration of refugees, which resulted in 90 per cent expressing their willingness to return. But the refugees seemed confused. Did registration mean volunteering for repatriation? UNHCR did not want to examine their willingness any closer. A subsequent awareness survey by Médecins Sans Frontières, Save the Children Fund, Oxfam, International Islamic Relief Organization and Terres des Hommes–Netherlands showed that a mere 9 per cent of the interviewed refugees were willing to return because they considered Burma to be safe at that moment. The survey also showed that 65 per cent of the refugees interviewed (412 families) were not aware they could refuse repatriation. Of the 61 per cent who said they had concerns regarding repatriation, 52 per cent stated that they could express these concerns, but 49 per cent thought it was too dangerous to talk. A majority of 75 per cent said they would eventually return once the situation in Burma improved.

Although UNHCR admits that the situation in Burma has not totally changed, it believes the refugees can return in safety and dignity. UNHCR believes that the Rohingyas are no longer discriminated against, that their fears, once well founded, are no longer justified, and that the UNHCR can effectively monitor their return. Other international human rights authorities, including the UN Special Rapporteur on Burma, Amnesty International and Human Rights Watch disagree with UNHCR's assessment. Their reports list continued human rights abuses, including forced labour and restrictions on freedom of movement. The UN Rapporteur reports that the Rohingyas face exceptional discrimination. Whether this caused the fresh influx of another 5,500 to 10,000 Rohingyas by May 1996 is yet unclear. The Bangladesh government denies access to the new arrivals who, out of fear, hide in the mountains or disperse among the local population.

Under these circumstances it remains questionable whether UNHCR can continue to repatriate the Rohingya refugees without severely infringing their freedom of choice.

Jeroen Jansen, Dutch MSF Project Manager, Asia

THE STAKES OF REPATRIATION

In an absolute sense, repatriation of refugees is probably the best solution, since their indefinite maintenance in camps is neither humanely acceptable nor politically desirable. This is of course assuming that the situation in the country of origin permits their repatriation, and that the international community guarantees that they return voluntarily. Along with the international system of protection for refugees, the basic principles for repatriation were conceived in the essentially European context of the Cold War following the Second World War. At that time, it was a question of the return of millions of persons displaced by the war to their countries of origin. After several false starts, the West defied Moscow and reasserted the policy of free choice by insisting upon the voluntary nature of repatriation, even of Soviet citizens, and by enlisting the protection of the international community. Traditionally, the UNHCR supported the notion of free and individual consent to repatriation and took no part in effecting repatriation unless it felt assured of the profound and lasting change of those circumstances that initially inspired exile. Thus it could facilitate the return of refugees in dignity and safety by negotiating agreements between the country of origin and the host country which would grant a minimum of guarantees to returning populations.

At the beginning of the 1990s, this traditional policy began to evolve. It would appear that priorities underwent a significant change of order: in the past several years, it seems that political factors have taken an upper hand to the once-essential principle of free and individual consent to repatriation. The 1993 report of the UNHCR working group on international protection emphasised that the voluntary nature of a return must be weighed

Plate 4.2
Refugees who have
fled the fighting in
and around Kabul in
a makeshift desert
camp outside
Jalalabad in eastern
Afghanistan.
Photograph: Benno
Neeleman

against the guarantees of security. Moreover, the rule of profound and lasting change has become subject to increasingly loose interpretation. The repatriation programme of the Rohingya refugees to Burma is a case in point: from July 1994, with the end of individual interviews and the inception of mass registration of refugees, the rule of individual consent was abandoned in favour of rather vague and general considerations regarding the political evolution of the Burmese regime.

Generally speaking, refugees are increasingly pressured to return to their home countries, and these repatriation programmes are not always conducted with respect for the professed principles of the international community. Some recent repatriations (Sri Lanka, Burma, Rwanda) have been carried out with a rather lax regard for real change in those conditions (conflict, repression, violence) that forced the refugees into exile and with little attention paid to the principle of voluntary return. It is as though the UNHCR has somewhat abandoned the question of free choice and the rule of profound and lasting change in favour of a vague notion of safe return or return in dignity and safety. To replace the notion of voluntary repatriation with that of a safe return is to replace the refugee's individual judgement with the discretion of the UNHCR or the states involved.

Plate 4.3
Rwandan refugees at a border crossing with neighbouring Zaire. Photograph: Klaas Fopma

This overruling of the will of refugees is all the more disquieting in that it increases the risk of pressure. In a context where repatriation is increasingly considered the only solution by the western countries that are the major contributors to UNHCR's budget, there is a variety of methods – for example, reduced food rations, or coercion, or intensified propaganda in the camps – by which pressure can be brought to bear upon those who are totally dependent upon international aid and the goodwill of host countries.

PREVENTIVE POLICIES

Apart from repatriating refugees to their countries of origin, the international community seeks to prevent new refugee problems. Preventing refugee situations arising poses the – political – problem of the attitude of the international community in the face of the repressive regimes and internal conflicts that provoke the major movements of refugees and displaced persons. The end of the Cold War gave birth to the ephemeral concept of a consensus of what is unacceptable and the illusion that massive violations of human rights within a country would no longer be tolerated.

Simultaneously, the UN, long paralysed by the East–West confrontation, became the recipient of these hopes and seemed, for a time, capable of playing a major peace-keeping role.

The euphoria, however, was short-lived. Events of the past few years have revealed the limitation of the speeches about intervention, and the problems of making peace-keeping missions effective underline the difficulties of international intervention in situations of internal crisis. Demands for protection far outnumber offers, and the highly selective international community is motivated to action largely by political interests, media visibility and public opinion. Preventive policies, the latest topic in international discussions, are actually mainly reactive, often late, and essentially defensive. Far from treating the causes of refugee exodus, western countries rather attempt to avoid the consequences. Incapable of finding a remedy to the source of the problem, they seek to stem the tide of flight.

The example of Iraq in the spring of 1991 is a good illustration of the international community's concern to avoid any new refugee problem, even at the price of providing a temporary protection for people repatriated to their own country. Coalition forces remained on guard before Iraq – though the country was already defeated and placed under international surveillance – during the bloody repression of Shi'ite and Kurd uprisings. But the spectacle of an entire people spilling over the frontiers of neighbouring countries and off the television screens finally provoked an international reaction. In the guise of an effective humanitarian enterprise, Operation Provide Comfort was geared to convince the suffering Kurds that they should leave the Turkish border areas and go back home in exchange for the offer of temporary protection and humanitarian aid in the north of Iraq.

The international reaction to the Kurdish exodus is perhaps the most clear-cut example of the new containment policy – or even push-back – based on the triad of repatriation, security zones and humanitarian assistance. It was as though this policy was designed to force refugee camps back into countries in crisis, in zones theoretically protected by an international presence and, in principle, supplied by aid convoys. Thus the mandate of the UNHCR has been extended *de facto* so that it could intervene in war-torn countries so as to encourage the process of repatriation and to assist locally displaced populations that threaten to spill over international borders. Modification of statistical criteria published in the 1995 UNHCR report is indicative of this evolution: listed as persons of concern for the UNHCR are not only refugees but also repatriated populations and over 5 million displaced persons.

This new policy has become a general one from Iraq to ex-Yugoslavia. Peace-keeping operations, despite their mixed results, have familiarised UN organisations with the concept of intervention in countries in conflict. And since 1989, the increase in assistance programmes based upon the idea of negotiated access, such as Operation Lifeline Sudan, have made this kind

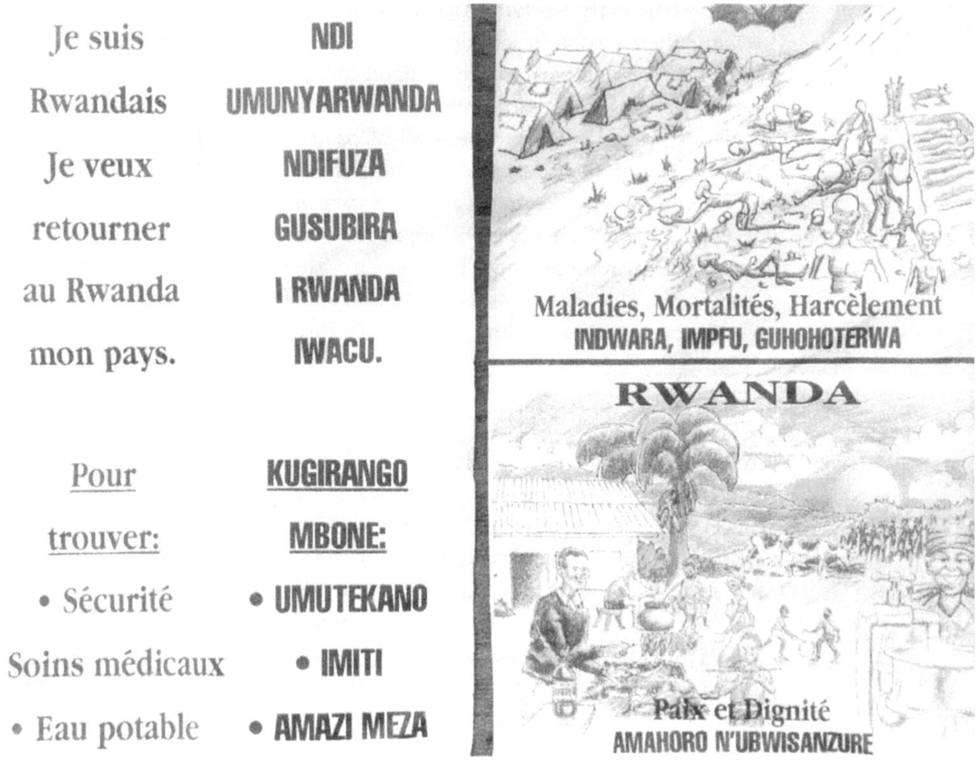

Je suis   NDI

Rwandais   UMUNYARWANDA

Je veux   NDIFUZA

retourner   GUSUBIRA

au Rwanda   I RWANDA

mon pays.   IWACU.

Pour   KUGIRANGO

trouver: MBONE:

• Sécurité   • UMUTEKANO

Soins médicaux   • IMITI

• Eau potable   • AMAZI MEZA

Maladies, Mortalités, Harcèlement
INDWARA, IMPFU, GUHOHOTERWA

RWANDA

Paix et Dignité
AMAHORO N'UBWISANZURE

of intervention in zones of contested sovereignty relatively common. Humanitarian aid is no longer simply distributed in refugee camps in peripheral areas of conflict, it is increasingly delivered to the very heart of combat zones in the countries in crisis, to help displaced populations. But this internalisation of international assistance means in fact the replacement of the concept of humanitarian sanctuary with that of the security zone, a factor which is more often than not detrimental to protection.

The result of military-cum-humanitarian interventions is in this respect disturbing. All the resolutions adopted by the UN Security Council in the past few years have concentrated on protecting aid operations but have neglected the protection of civilian populations. During the war in Bosnia, the Blue Helmets of the UN provided protection for the relief convoys but did nothing whatsoever to end the massacres and ethnic cleansing. They proved incapable of defending the security zones they had themselves set up, and finally had to abandon them to the Serbs. The mediocre results of UN interventions in the past several years underline the fact that it is difficult to protect displaced populations which flee for the same reasons

Plate 4.4
Pamphlets aimed at encouraging refugee repatriation dropped from United Nations aircraft over camps in Goma, Zaire.
Left panel: I am Rwandan. I want to return to my country, Rwanda, to find security, medical treatment and drinking water. *Top right panel*: Disease, Death, Harassment. *Bottom right panel*: Rwanda: Peace and Dignity. Photograph: Esmeralda de Vries

as refugees do, but which remain within their country's borders and consequently are not recognized by the international community.

It is particularly worrying that preventive policies to stop refugees from crossing their own frontiers deny them the benefits of a system of international protection based precisely upon the notion of crossing international borders. It is as if potential refugees were being relegated to the status of displaced persons, deprived of any real protection and offered only uncertain assistance which depends on conditions of access. This is all the more disquieting because western countries which provide the quasi-total of UNHCR's budget are often willing to finance aid programmes in countries in conflict merely to pay lip service to their obligations to the 1951 Convention. The creation of security zones and aid programmes cannot serve as a pretext to refuse asylum to seriously threatened populations.

THE QUESTION OF ASYLUM

Confronted with a considerable increase in requests for asylum, western countries have in recent years dramatically modified their asylum policies. The effects of this change were quickly felt, for, while the number of demands for asylum rose from 70,000 in 1983 to close to 700,000 in 1992, they were reduced to less than 300,000 by 1995. At a time of increasing crises and in the context of restrictive immigration policies, a demand for asylum provides the only access to the industrialised countries. The dramatic drop in the number of such demands in recent years indicates the effectiveness of methods of dissuasion, using administrative and policing measures. Facing a veritable obstacle course, very few asylum-seekers make it to the portals of Europe, and when they knock at her doors, they are frequently met with flat refusal and are shunted off to the last safe third country of their transit. The rules have multiplied, legislation has become more restrictive, and constitutions have even been revised in order to legitimise the refusal of access to these people to a Europe which increasingly resembles a fortress. This hardening of policy, which is taking on the aspect of a lock-out, is justified by an obsession with economic refugees; nevertheless, it brings into question the very right to asylum that is a founding principle of so many European democracies.

Simultaneously, the percentage of people accorded refugee status has plummeted from 42 per cent in 1984 to less than 10 per cent in 1995. The harmonisation of refugee policies within the European Union is based upon the lowest common denominator, and the Fifteen are increasingly restrictive in their interpretation of the concept of persecution, contrary to the spirit and the letter of the 1951 Convention.

Thus on 23 November 1995, European Justice and Interior Ministers concurred that only people persecuted by a state could be defined as refugees. This new definition refuses international protection to the victims of

extremist movements – Algerians threatened by the GIA (a splinter group of the Front Islamique du Salut), for example – and to nationals of states that have virtually disintegrated – like the Somalians and Liberians who are at the mercy of armed bands – not to mention the Bosnians, who have been flushed out by ethnic cleansing and who unquestionably meet the

Plate 4.5 Bosnian refugees at a checkpoint on their way to Tuzla. Photograph: Christian Jungeblodt (Signum)

1951 Convention definition of refugees. This disastrous decision is evidence of a will to be rid of the refugee burden while turning a blind eye to the evolution of crises and the distress of victims. This is how European countries seek to unload the obligations undertaken when they signed the 1951 Convention. Under judicial cover, the refugee question has always been answered with political and ideological considerations. Today the spectre of mass immigration has led European countries to refuse protection to nationals of countries in crisis, and the risk is that this tendency of western countries may become generalised and spread to the entire planet.

There is a definite correlation between the drastic reduction in the number of people European countries have accorded the status of refugee and the number of *de facto* refugees – people who are not recognised as refugees but cannot be sent back to their homes in turmoil-stricken countries. This development is all the more apparent since, for the first time since the Second World War, the problem of war refugees, once restricted to the south, has become a European problem. Confronted with a massive exodus of refugees, European countries have created temporary protection procedures which allow them to accept refugees from ex-Yugoslavia without making any commitments as to their status. Clearly, the great exoduses caused by conflicts do not fit with policies of planned reduction, and European countries fear an unexpected extension of their international obligations. So these new procedures grant neither the same guarantees nor the same rights, and these refugees are merely tolerated; the future of their temporary and precarious status is left entirely to the discretion of public powers.

NEED FOR A REFUGEE POLICY

Faced with the greatest exodus of war refugees since the Second World War, European countries cannot continue to treat the problem with stop-gap humanitarian measures; they must define a clear policy and assume their obligations according to the 1951 Convention. Any liberal refugee policy rests on a combination of an open-door policy, founded on respect for the right to asylum and a discriminating hearing process that allows for a selection between refugees and immigrants. The primary obligation of any state is to preserve the right to asylum and not to allow the fear of migrating masses to stifle those values basic to all democratic states. Its second obligation is to effect just, rapid and adequate procedures to determine the status of war refugees. In an atmosphere of narrow-mindedness with regard to identity, of increasingly restrictive immigration policies, and of increased population movements, criteria must be established to protect those who are most threatened. And its third obligation is to treat refugees and displaced persons humanely, and to guarantee them aid and protection.

Definition of coherent policy is all the more necessary because the attitude of democratic countries regarding the defence of human rights and respect for international norms has a determining influence upon that of countries which will receive the main flows of refugees. At a time when the right of asylum seems to be visibly shrinking, the risk is great that the European example will serve as a pretext to other countries which would institute closed-door policies of ostracism. But quite apart from the repercussions of changes in European policies upon those of potential host countries in the south, the attitude of northern countries in general considerably affects policy decisions all over the globe. Western countries, which finance in large part the activities of the UNHCR, have considerable influence in the current redefinition of institutional priorities of prevention and repatriation.

Dissuasive procedures adopted by European countries, repatriation operations set up by host countries and greater preventive measures in countries in conflict are all simply aspects of an overall policy designed to maintain potential refugees in their countries of origin and, failing that, to get rid of the refugee burden by keeping them at a distance or by sending them home. In the north as well as the south, one hears the watchwords of the same logic: humane deterrence, safe country, safe return or humanitarian aid. It is, indeed, a broad-minded humanitarian who allows states to slither away from the problem while they profess the very best of intentions. Thus the humanitarian alibi allows western countries to manage the problem of war refugees without actually granting them refugee status.

By the same token, in countries in conflict, humanitarian aid lends credibility to the idea of an international commitment to displaced populations while it limits the risk of refugees spilling over international borders. This tendency towards favouring a humanitarian whole is disturbing, because it frequently turns out to be detrimental to protection. Given the will of western countries to close the refugee chapter, the UNHCR faces an increasingly difficult task in maintaining a balance between solving the refugee problem and ensuring the protection of refugees. In this unfavourable context, humanitarian organisations have an essential role to play in reminding the UNHCR of the principles it must respect, and in reminding states of their international obligations.

5

FALLING THROUGH THE NET
OUTCAST AND MARGINALISED POPULATIONS

Julia Groenewold and Stephan van Praet

•

INTRODUCTION

Traditionally, most frontline relief organisations have focused on the plight of civilians caught up in wars and natural disasters. Yet perceptions are changing. More so than before, the general public in Europe and North America is questioning the reasoning behind the efforts of high-profile aid organisations to help afflicted populations abroad while people in their own backyard are being neglected. Many westerners, too, would be shocked to discover how similar basic conditions in the inner cities of America, the north of England or the back streets of southern Italy often are to those in the Third World.

International humanitarian organisations can no longer ignore the fate of the growing numbers who are suffering in the shadows of the so-called developed world. The homeless, the illegal immigrants, the inhabitants of *favelas* and shanty towns, those diagnosed HIV-positive, ethnic minorities – there is a growing list of social groups which have fallen through the safety net of social services and are in urgent need of some form of minimum protection and assistance. As a result, key medical and relief organisations, including Médecins Sans Frontières, are turning increasingly towards trying to improve such conditions, or at least draw attention to them. Marginalised by media headlines of catastrophes all over the globe, the problems of our neighbours often remain unknown. The invisible suffering of elderly people who can no longer survive in a decent manner is no less important than the impact of an exotic emergency – it simply looks less dramatic.

Throughout the world, populations face the indiscriminate violence of nature, but suffering is not limited to such natural disasters; societies that refuse to care for or neglect their citizens in need cause them much unnecessary suffering. Economic realities affect some more than others and deprive entire communities of social benefits that are taken for granted by other segments of society. Even in countries with well-developed social security systems, such as Belgium and France, the number of people without social security is growing. Many have given up trying to negotiate the complex

administrative procedures involved in collecting benefits. Others are too intimidated by the anonymity of the system to make the attempt. The impersonal urban environment renders institutions and individuals indifferent towards their fellow citizens. When neither families nor schools are able to provide a minimum level of socialisation, the law of the jungle takes over and 'survival of the fittest' takes on a very real meaning. The victims of these social disasters are in danger. Their protection needs to be high on the agenda of human rights organisations and humanitarian agencies.

Relief agencies attempt to alleviate individual suffering. However, it is the cumulative effect of individual human tragedies that has created today's social reality. Although charity on its own can help some individuals, it is not sufficient to transform the mechanisms responsible for that reality. Relief activities in developed countries not only involve direct attention to individuals in need; they attempt to put pressure on local authorities and decision-makers to face up to their responsibilities towards growing numbers of the socially excluded.

This chapter tackles the issue of social exclusion in a developed environment. The role of humanitarian agencies in such an environment is open to debate. Should they just be providers of services on an international scale aimed at fighting the negative aspects of world-wide economic problems,

Plate 5.1
James has been living with HIV since 1985. He eventually moved into the Ambassador Hotel in San Francisco, CA, which has served as a refuge for HIV/AIDS patients since 1990. Photograph: Paul Fusko (Magnum)

or should they sound the alarm and act as witnesses to the growing cancers within our egoistic societies, with the goal that governments will assume responsibility? The cases described here are based on some of the field experiences of MSF. Unfortunately, there are numerous other cities, countries and social groups that could have been included as well.

OUTCASTS IN THE SHADOW OF SKYSCRAPERS

While many people stride forward along the path of economic progress, an increasing number are left on the sidelines. They represent a variety of different groups, sharing a common fate forced on them by the mutually reinforcing realities of economic marginalisation, limitations relating to work skills and abilities, poor education, racial discrimination and isolation from the welfare system.

MSF and other western frontline aid agencies are becoming increasingly involved with people in need in their own societies. Indeed, humanitarian organisations which base their policies and actions on respect for medical ethics, human rights and international humanitarian laws must address exclusion everywhere. In war zones and crisis areas they fight for the respect of the universally declared right to a standard of living adequate for the health and well-being of every person, including food, clothing, housing and medical care and necessary social services. They certainly cannot tolerate violations of such rights in their own backyards in societies where governments have the means to protect and care for their citizens. In order to help excluded groups within their own societies, western medical groups and other humanitarian organisations offer medical services and try to draw public attention to the problems.

Medical care and social exclusion

Increasing restrictions on access to health services bring medical relief agencies to the forefront of the issues related to social exclusion. In most countries, people who fall outside the medical insurance system have to resort to the emergency services of a public hospital. In the United States alone, the use of such services affects at least 20 million people. Emergency care has to fulfil *de facto* the function of primary health care. This has led to an impossible situation. The emergency services are overstretched and do not have sufficient time to deal adequately with patients. There is no follow-up and no preventive care – emergency wards are neither equipped nor intended to deal with the medical problems presented by the outcasts who are forced to turn to them for help.

There has been a striking reappearance in the health statistics of the most developed nations of diseases that have a poverty stigma, like tuberculosis and scabies. Clearly, a gap in health services has allowed the renewed outbreak of these diseases. In Paris, a system of nursing beds was set up in March 1994, providing treatment by nurses rather than doctors. With programmes such as these, a social apartheid has emerged in health care, where different levels of society are offered different levels of care. Adequate health care is becoming a luxury rather than a basic human right in more and more countries.

Immigration – at the heart of social exclusion

It is widely believed that illegal immigrants represent the core of the social exclusion problem. The term 'illegal' masks the complex reality of many different categories of people without official residence papers. The International Labour Organisation estimates that there are 30 million such illegal immigrants world-wide, 4.5 million in the USA alone. As the numbers rise, the European Union is tightening up its external borders and the USA has been accused of building a militarised 'tortilla curtain' along its southern border in an effort to prevent massive south–north migration. Tighter borders combined with new legislation limiting access to social services for illegal immigrants is making their plight even worse.

Newly proposed immigration laws in Europe and the USA would oblige school principals and hospital directors to check on the immigration status of students and patients. Such laws not only restrict the rights of immigrants, but also make it illegal for people to help illegal immigrants to survive. Surely we are approaching the limits of the inhumane when supposedly free and democratic societies assume the social paranoia of authoritarian regimes. When society denies children education and refuses the sick essential health care, it can only be sowing the seeds of enormous social problems in the future. Medical groups are attempting to bridge the health services gap, but while they can provide essential care, they cannot remedy the ills brought on by a society that refuses to take care of all its people.

Medical aid in Holland

'It was never a deliberate choice. We never said: "From now on we will treat people without insurance. They just came,"' explains physician Markus Kruyswijk who cooperates with De Witte Jas (the White Coat), a Dutch organisation that provides medical care for patients without health insurance in Amsterdam. Not only has the number of patients of De Witte Jas grown considerably in the ten years of its existence, but the social composition of these patients has changed dramatically. In most cases they are immigrants who came to Holland for work. Other clients are rejected asylum-seekers, homeless people, young wanderers, drug addicts, alcoholics and psychiatric patients.

A supporter of the organisation says, 'De Witte Jas is the nearest thing

Plate 5.2   A homeless couple begging on the winter streets of New York City. Photograph: Thomas Hoepker
(Magnum)

to development aid given in a developed country. Our clients who come from Third World countries have in Holland returned to a Third World situation.' Officially, illegal immigrants have access to health care in Holland, but only if they pay for it. For most of them, low-paid workers or unemployed, this obviously leads to enormous financial problems. A small number of general practitioners and organisations such as De Witte Jas provide primary health care, but this kind of small-scale care is obviously not ideal. However, budget cuts in the health sector and government regulations that restrict official access to health care to life-threatening situations alone, make this voluntary assistance more and more necessary – and more problematic. A real problem arises when the patient needs to be referred to a hospital. In Amsterdam, hospitals have always been willing to make exceptions for patients who have no insurance. Although the hospitals have received some reimbursement from social security and the National Health Service, they have incurred enormous financial losses. New government regulations now place very restrictive conditions on such reimbursements. Consequently, hospitals have become extremely reluctant to provide non-emergency care to illegal immigrants. 'So, if a patient without health insurance has cancer, help will only be given when his situation has become life-threatening. That is when the cancer has spread through the body, when it is too late,' said nurse Eef van Boeckel in an interview with a Dutch magazine.

The Dutch government fears that if illegal immigrants are allowed access to social and health services they might get the impression that they are tolerated. Leo Adriaenssen, a physician who collaborates on an informal basis with De Witte Jas, does not believe that illegal immigrants will leave because they cannot get health care. 'Large numbers of people from poor and war-torn countries will continue to come to the rich western countries to try to establish a better life. It is unacceptable that in Holland seriously ill patients remain untreated and have to rely on charities and private persons to save their lives. And it is unethical and inhumane that the denial of medical care is used as political pressure to make illegal immigrants leave the country. The authorities can't enforce legal expulsion from Holland and therefore have turned to an indirect physical threat as a solution,' he said.

Mounting obstacles in the United States

In California, physicians are confronted with similar ethical and moral problems concerning illegal immigrants. In 1870 Senator William Stewart of Nevada said, 'I am opposed to Asiatics being brought here, but while they are here it is our duty to protect them from barbarous and cruel laws that place upon them unjust and cruel burdens.' At the time, waves of attacks on Chinese immigrants swept through California, and the California legislature imposed a series of wildly discriminatory taxes on the Chinese. Nearly

125 years later, in November 1994, California voters approved Proposition 187, which requires publicly funded health care facilities to deny non-emergency care to illegal immigrants and to report them to government officials. Since the passing of this proposition, two illegal immigrants have died, allegedly because they had delayed seeking medical care for fear of deportation.

'Proposition 187 is a fiscal measure, not a health measure. If physicians report illegal immigrants to help enforce the law and balance the state budget, why not also identify tax evaders, traffic-ticket scofflaws, or parents who fail to pay child support?' write Tal Ann Ziv, BS, and Bernard Lo, MD, of the University of California in the *New England Journal of Medicine* of April 1995. They further state that if doctors cooperate with Proposition 187, they forgo the ethical ideal that patients' medical needs should be attended to without regard to their social, political or citizenship status. Traditionally physicians have had a special humanitarian role, serving even unpopular or antisocial patients, including people charged with crimes. Other opponents of Proposition 187 contend that immigration laws should be enforced at the border and in the workplace, not in hospitals and clinics.

Supporters, however, argue that those who are ill will return to their native countries for medical care. They think that welfare, medical and educational benefits are the magnets that draw these illegal aliens across their borders. 'It is equally plausible that people will deny their illness, try home remedies, obtain medications from friends, and delay seeking care, thereby worsening their medical condition and potentially threatening public health', maintain Tal Ann Ziv and Bernard Lo. They propose that physicians should, at a minimum, oppose policies regarded as unethical and attempt to overturn or mitigate them through public education and political action. Organised medical groups should oppose similar measures in other states and in Congress. The American Medical Association has already denounced Proposition 187 and opposes any federal regulations that would require physicians to determine the immigration status of their patients.

MSF'S OPERATIONS AT HOME

While traditionally focusing on aiding populations in war zones and conflict areas, MSF has become active in assisting excluded people within its home base, western societies.

France

In France, for example, MSF has often vehemently denounced government policies denying fundamental care to some French nationals and immigrants.

In April 1996, MSF condemned a draft immigration bill that threatens the rights of foreign nationals living in France, especially their right to health care. The bill states that providing accommodation to a foreign national without a residence permit is a criminal act. The host is then guilty of aiding and abetting illegal residence in France. Furthermore, the law applies even if the host is coming to the aid of a foreign national who is ill. In press statements, MSF has pointed out that this bill excludes illegal immigrants from the health-care system. Furthermore, any attempt by relief workers to obtain information from the sick individuals could result in danger to family, friends and anyone else involved.

This implies a *de facto* denial of care. MSF states that the law that allows foreign nationals without residence permits access to the country's health care must be respected. And this law has to be protected from any kind of formal or informal (through constraining administrative procedures) abrogation. The medical welfare machinery must foot the bill, the patient must be visible, and there must be no threat to his family and friends. A patient, whether a Frenchman or a foreigner, whether legal or illegal, must have treatment, a recent press statement said.

Since 1989, MSF has fought for universal medical care in France. In cities like Paris, Marseilles and Lyons, the organisation has opened centres for medical and social care. These centres help jobless people get their benefit entitlements, people on income support and vulnerable foreigners. They give care to wanderers who have no idea that they are ill and to homeless people who are incapable of filling out an application, let alone standing up for their rights. They try to save people such as one illegal immigrant who was dismissed from hospital with the warning that he would die if he did not take his insulin injections twice a day. Without papers or money, he could not buy the medicine and lapsed into a coma. He came out of it only ten days later.

Apart from this practical assistance, MSF has, for example, lobbied tirelessly with deputies to vote in favour of a law (l'article 186 du Code de la famille et de l'aide sociale) guaranteeing medical aid to every foreigner living in France. MSF has extensive experience in treating patients with severe medical conditions who have been categorically refused such treatment by doctors, administrators and social workers. MSF doctors and social assistants put together dossiers and other material showing that, without the proper papers, one could die at the entrance of a hospital in France. These dossiers, countless phone calls, persistence and tenacity in the end resulted in moving the mountain: the law was accepted in July 1992.

Belgium

In Belgium, MSF has been providing medical and social support to both the homeless and illegal immigrants since 1989. Medical health care and social welfare advice is given in health units throughout the country, as well as via a mobile clinic in Brussels. Well aware of the increasing numbers excluded

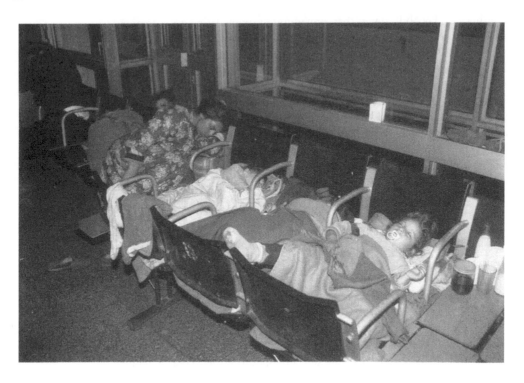

Plate 5.3
Gypsies find a
quiet place to rest
in Brussels.
Photograph: Roger
Job

from medical care, even in emergencies, and recent ominous proposals for even more restrictive legislation concerning immigrants, MSF has adopted a policy to intervene actively on behalf of immigrants and to lobby the government and its institutions to take a more humanitarian approach.

'Europe is closing its borders to immigrants who arrive without papers, without any social security. But they still arrive and MSF wants to help them,' said Ines, a medical doctor in an interview with *Le Soir Illustré* (Marcel Leroy). 'And we will inform the public about the work we are doing, because our relief activities must not become the excuse that allows authorities to evade their responsibilities,' she added. Ines works in the MSF clinic bus parked at the railway station in Brussels.

Ines has helped and witnessed a wide variety of patients on the clinic bus. Once, she helped a man who was afraid of dying of a heart attack. He was about 60 years old and clearly suffered from poverty and loneliness. Squeezed together on a bench one might find a few African immigrants, a pensioner from Brussels, a young Italian, an old lady and a man who sleeps in the railway station. Few who visit the clinic bus have any idea there are social services open to them, but sometimes MSF staff can refer them to available services. On one occasion, the clinic staff tried to help the Zairean mother of two who was eight and a half months pregnant. She was in Belgium illegally and was completely exhausted and demoralised when she

turned up at the MSF clinic. Her husband had just died of liver cancer, so the Belgian public centre for social welfare had suspended support for the family, and they officially had to leave the country. The woman could not pay the rent any more and had to leave her house. Friends took her in with her two children. She had not had a medical consultation for months.

Plate 5.4
Homeless men in Brussels.
Photograph: Roger Job

Plate 5.5
Homeless in
Brussels.
Photograph: Roger
Job

SOCIAL EXCLUSION AROUND THE WORLD.

Through its programmes in the *favelas* of Rio, in the slums in Nairobi, the garbage dumps of Guatemala City and with gypsy communities in Romania, MSF encounters and struggles against the sordid reality of exclusion that exists in the most diverse social environments. To some it may seem a hopeless objective to become involved in alleviating social exclusion all over the world. As aid agencies and human rights groups are only too aware, the world is besieged with numerous different groups, all encountering varied problems in incomparable circumstances and contexts. Where to begin and where to end?

Rio de Janeiro: the favelas *under alternative protection*

Rio de Janeiro is an exotic Brazilian megacity of dream beaches, carnival rhythms; a multicultural melting pot where every modern product is available. But this is not true for everyone. Certainly not the street kids or those living in the growing *favelas*, the shanty towns controlled by the underground. In 1993, MSF started a programme in Rio that sought to provide medical, psychological and social services to street children and young single mothers, in association with a number of local organisations. The programme further evolved in 1994 in response to the appalling living conditions among the 12,000 inhabitants of the *favela* Vigario Geral. Drinking-water was hard to find, there were no health services and the schools were closed. MSF has been providing free medical care for children and is participating in setting up a cooperative which will offer medical services for adult patients. Water supplies have been improved and a sewage treatment system installed.

As in other *favelas*, the drug economy produces extreme violence and tough retaliation by government forces. MSF's programme had to grapple with the complete lack of social and physical protection for those working in the *favelas* and the vacuum that isolates the population from state authorities, which are perceived as corrupt, oppressive and violent intruders into their lives. MSF took over medical services that would normally be the responsibility of the state, and had to make tacit agreements with local criminal gangs for protection. The government considers such communities suspect, and the MSF centre and its staff represent targets in government raids.

Living in the shadow of modern shopping centres, the *favelas* are a fertile breeding ground for criminal organisations. They suffer both an economic and a social exclusion from a society that treats them as enemies and seems to be only interested in repressing them. No wonder then that they fall easily into the protective arms of the outlaws.

Romania: gypsy minorities treated as second-class citizens

The vulnerability of ethnic minorities in Romania and other European countries is well illustrated by the experience of the gypsy (roma) populations. They are pushed onto the margins of society as the result of structural discrimination based on a history of systematic neglect by local and government authorities. 'They are lazy and cannot adapt to our society,' the *gadje* (non-gypsies) say. 'We are suffering from structural discrimination and have been systematically neglected by local and government authorities,' the gypsies say. Consequently, a mutual suspicion dominates the relations between gypsies and *gadje* and sustainable social contacts are extremely difficult to attain. Any efforts to improve the living conditions of the gypsies will take a lot of patience and time.

Nevertheless, MSF has tried to help some gypsy groups in Romania with programmes in small communities. Pata Rat, a settlement of garbage collectors at the end of the N 8 bus line from Cluj, is one of them. The gypsies, who have christened their settlement 'Dallas', live in an isolated spot in run-down houses. Families of eight to 15 members survive in poorly furnished houses of one or two rooms built out of cardboard, pieces of wood and other rejected materials. The houses often have no windows, proper heating systems do not exist and there are no sanitation facilities. Most of these families have to live on the salary of one family member, if one has

Life in a favela

On their way home at night, they clamber up unlit stairways built without a permit and cross a narrow walkway over a couple of walls and a railway line to get from civilisation and established order to their humble homes in an isolated enclave run by the local drug dealers.

Home is a *favela* of Rio de Janeiro.

The violation of fundamental human rights is a fact of everyday life for these people. They live in areas outside government control. Theirs is not an economic or social exclusion, since a good number of them earn an honest living. They are domestic servants in Rio's residential areas, skilled workers, watchmen in buildings or private homes or the employees of security firms.

What they don't have are proper homes, which is why thousands of them have invaded the hillsides of Rio de Janeiro. They are excluded because the Brazilian government has never really wanted to acknowledge the existence of the *favelas*. There is no police station, one solitary school, no health centre, a single telephone and none of the infrastructure needed to install more.

An alternative state has grown up as a result. The drug dealers, who were initially interested only in money, have taken on a political and social function. Handling a steady stream of requests for help from the *favelas*, the local gang bosses are performing key roles that only the state could perform in their stead.

A despairing mother has no way of paying for the burial of her daughter run over by a bus. She has to get the body from the hospital to the cemetery, buy a coffin and flowers, and get herself and the family to the distant cemetery. The drug dealers hire a minibus to take everyone to the cemetery, buying flowers and a coffin to give the child a decent burial.

Maria falls ill and urgently needs hospital treatment. It is Saturday night and no ambulance will enter the *favela*. The dealers immediately provide a van and driver to take the poor woman to hospital.

Thus the state finds itself trying to unseat a rival authority, but does not have the political will to replace it. The authorities appear to want to root out the evil, without leaving anything in its place.

What does MSF do in a situation like this? How can it work in a *favela* rife with drug dealers, who often enjoy popular backing because they are the only source of help in times of need?

In such cases, assistance is a political act. The NGO [non-governmental organisation] can work with the people and provide a service that the state does not provide and that the drug dealers provide only in emergencies, thus filling a void and limiting the power of the dealers. The dealers need the people and their support to be able to operate in this enclave. An MSF health centre that enjoys the backing of the community because of the number of health problems it resolves cannot be rejected by the traffickers and is not (so far at least) perceived as a threat.

We take no political stance, but our action is political. If the state and NGOs were to invade this *favela* with education, health, sanitation, recreational and training projects, the power of the drug trade would be greatly reduced. The people would have no further need of the traffickers' help, young people in the *favelas* would have genuine alternatives to crime, and the outrageous gulf between the Brazil of Ipanema and that of Vigario Geral and the 600 other *favelas* would be bridged.

Thousands of children visit the health centre and a health cooperative has been set up for adults, with no input from the drugs trade. These are political acts. They help these people take a step towards citizenship. The step may be tiny, but it is the first in a long march.

Michel Lotrowska, Belgian
MSF Programme Coordinator, Brazil

been fortunate enough to find employment. The priority of this MSF programme was to improve the health status of the gypsies in this community. In order to achieve this, some other aspects of their social context, such as children's education, had to be tackled first. After the fall of the Ceausescu regime, during which all children were obliged to attend school, very few gypsy children continued to go to classes. The parents offered many excuses, including a lack of suitable clothes or shoes, poor hygiene, no money to buy books and the necessity for children to work or to look after their younger siblings. Lack of education obviously limits their chances for employment, keeps them ignorant of administrative procedures and alienates them from public services. And in the already difficult circumstances that the gypsies live, it doesn't take a lot more to fall into the downward spiral of social exclusion. Therefore MSF, after gaining the confidence of the community and the relevant authorities, worked on convincing the gypsies to send their children to school. Furthermore, MSF tried to improve the water quality and helped the gypsies navigate the administrative procedures necessary to get proper housing and other social services.

*Protection for transnational minorities**

The countries of central and eastern Europe generally provide better legal protection for their minorities than the member states of the European Union. The new constitutions that were adopted after the fall of communism recognise a number of specific individual rights and freedoms for

---

* See Claire Auzias, *Les Tsiganes, ou le destin sauvage des Roms de l'est*, Ed. Michalon, Paris, 1995.

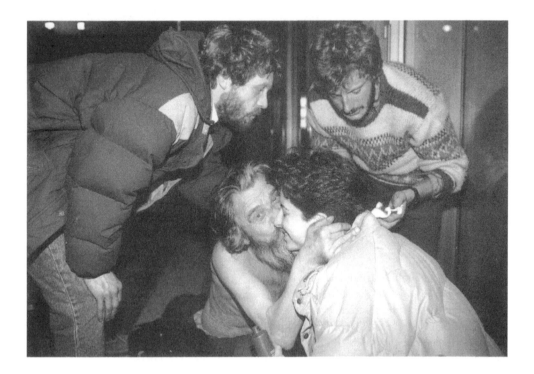

Plate 5.6
An MSF volunteer
visits gypsies in
Barcelona, Spain.
Photograph: MSF

members of ethnic minorities. However, as illustrated above, gypsies have more difficulties than other minority groups in dealing with the practical implications of acquiring such official protection and are often not accepted as full citizens in the countries concerned.

International law neither recognises nor defines the concept of such transnational minorities as gypsies, and whatever rights they have are therefore generally limited to what is offered by the legislation of the country in which they live. Collective rights for transnational minorities have not yet become a subject for widespread discussion. As a result, there is a major problem in regard to access to citizenship. In the Czech Republic, for example, thousands of gypsies have systematically been refused Czech nationality under a law passed in 1993 and are treated as stateless persons. They are extremely vulnerable, suffering social discrimination and constantly fearing expulsion. This new law aroused much international protest but that could not resolve the situation of these stateless persons.

Moscow: residence registration and red tape

In 1991, MSF initiated a project to take care of Moscow's socially excluded: its homeless population. According to the police, Moscow has around 300,000 non-registered residents. Russia's cumbersome and complex administrative system restricts the right of individuals to live where they choose. Every citizen is required to have a *propiska* (registration stamp) in his/her identity papers. The place of registration is usually the place of birth or work location. Big cities such as Moscow or St Petersburg are so-called 'closed cities', where the local authorities try to prevent new immigration. They therefore require re-registration every month.

In Moscow, it is illegal for anyone without the official registration stamp to remain in the capital for more than three months. Moreover, without registration, people have no legal status and lack access to free medical care and other social services. MSF's programme in Moscow is focused on prevention and health-care education for the homeless and seeks to improve their socio-economic conditions. In 1995 alone, about 25,000 free medical consultations and vaccinations were given in one fixed and two mobile medical centres. A referral system has been set up in collaboration with six hospitals. MSF gives advice on how to use the two public disinfection centres operating in Moscow and is providing assistance with the paperwork necessary to regulate their official status.

They assisted, for example, Vassili Andreevitch Savieliev, who is 91 and lived on the streets of Moscow. He was too old and too weak to defend his rights when his insane daughter kicked him out of their house. Without help he would never have been able to collect the necessary documents to be able to move into a home for the elderly. No civic administrative office would help him. MSF can help by giving people in these situations a Human Rights Certificate that gives basic identity information for those who have lost their official papers. Such certificates, which MSF draws up, provide a minimal form of protection for those subject to police raids. These initiatives are being developed in collaboration with numerous independent and state organisations.

Eighty per cent of the homeless in Moscow are Russian, 90 per cent are male and one in four is an ex-prisoner, like Aleksei Constantinovitch Ibragumov. Aleksei, a Muscovite born in 1960, was sentenced to 11 years in prison in 1983. While he was in jail his sister sold their five-room apartment. After his release in 1994, he found himself homeless and penniless in a world that had changed dramatically during his prison term. Under recent decrees by the Moscow authorities he should have been eligible to housing rights as an ex-Muscovite, but he had no idea how to find his way through the system to get his name on a waiting list to provide him with a roof over his head.

WHERE TO BEGIN AND WHERE TO END?

If you lift up the net of social protection, you will find that a wide range of groups and individuals have fallen through its holes for a variety of reasons. These include a lack of access to the social security system; failures

The times they are a-changin'

The following is a letter written by a 1990s relief volunteer to a friend and founding representative of MSF, who first joined the organisation nearly a quarter of a century ago.

Dear Alain:

Two months into my Moscow mission, I've eventually got round to writing: and I have to tell you, the problems you encountered 25 years ago were different.

You were never able to remain indifferent to the plight of refugees, displaced persons and the victims of war – or, later on, to ethnic cleansing. The people you came across during your missions were there as a result of some exceptional event (war, famine, epidemic or natural disaster). They had been forced to leave everything behind. You publicly condemned these horrors over and over again, in the hope that others would be stirred by your sense of shame.

You used to say that you would never cease to be amazed by the resourcefulness of those caught up in such situations: they never lost their grip – family, friends, acquaintances or clans all helped them rapidly rebuild a community they felt they belonged to. At first sight, you said, they had lost everything – and yet it was heart-warming to see life could resume its course, even in the camps. Their resources allowed them to pick themselves up, to retain their dignity, to prove they could survive.

Well, now you're working in the international office and I'm on my third mission. This time, I don't feel I'm at the ends of the earth – it's more like the end of the world. All those children in Manila, Lima, Bucharest or Rio, damaged physically and mentally by the rubbish they sift through, the homeless we pass at the exits of metro stations in Paris or Brussels, or their counterparts in New York or Moscow who take the slightest gesture as a hostile, threatening signal. All of them have become marginalised in their own societies – without any particular upheaval having occurred. You can find them in all the big cities, even the richest.

They have been rejected. Families have turned away, friends have sunk with them. They no longer have access to the welfare support they are theoretically entitled to. They've lost all their bearings: they don't feel they belong to a community any more. They have lost their sense of identity.

Social exclusion used to be a far-away phenomenon – but now it isn't just on our doorstep, it's actually moved indoors. Poverty has become the world's biggest epidemic. Some claim there is no remedy. The fact that poverty is spreading into the North doesn't make it any more unjust in itself, but it does illustrate the devastating speed of its progress. A lengthy illness, unemployment or excessive debt can push anyone over the edge. No one is immune.

These people have no stake in society. They can bring no economic or political pressure to bear. They are the victims of social breakdown and economic recession, but even more of a crisis in values like respect for the dignity of the person, respect for and acceptance of differences, and the sense of shared responsibility for those who are not competitive in society. These are the values underpinning the very concept of human rights. These rights are not abstract: they belong to real individuals who are ill or hungry, perhaps children. Such values also entail human duties, which are binding upon the rest of society.

If social exclusion is multi-faceted, then so too must be the search for solutions. Might combating poverty involve separating the right to an income from the obligation to work? Should the idea of universal allowances be encouraged? Does real protection for individuals depend upon moving away from the concept of social insurance to that of a debt on the part of those who are lucky enough not to need it?

These factors put the concept of humanitarian action into a new perspective: a humanitarian ideal which bears witness to our demands, including the right to a decent life for those who no longer count. A humanitarian ideal which reminds the countries of the North that human rights are being crushed underfoot there too. I'd like to finish this letter with a phrase by Jankelevitch to the effect that moral indignation is the sole vehicle capable of moving us from passive contemplation of inequality to a revolutionary rejection of the morally unacceptable.

Benoit Deneys, Belgian MSF Programme Manager

in the system to provide protection; malfunctions in the system itself; discrimination against one group or another on the basis of ethnicity; economic conditions; social and geographical isolation and even health problems.

MSF faces a major challenge in trying to tackle this huge problem. Assistance programmes on their own are not enough. Medical and humanitarian organisations must support initiatives in the area of legal protection and take an active part in the debate on ethical principles that will become the basis of policy for our societies in the future, at both the local and international level. If not, humanitarian agencies risk simply treating the symptoms of societal exclusion rather than helping cure the disease. When public policies, or a lack thereof, create humanitarian disasters, humanitarian organisations must speak up on behalf of the victims.

As humanitarians, we defend the rights of vulnerable populations. We have to convince the political decision-makers that humanitarian assistance alone will not prevent breaches in the protective dike of social security and the welfare system. Social injustice must be fought with policies that aim

Bangkok: AIDS patients easily excluded

The AIDS pandemic struck Thailand in the early 1990s, the first country in Southeast Asia to be confronted on a large scale with the enormous financial and human consequences of the disease. The Thai authorities, who had become aware of the seriousness of the problem in 1991, started a vigorous campaign aimed at information and prevention. It became obvious that those suffering from AIDS in the rapidly expanding capital of Bangkok would become a highly vulnerable group, isolated not only from economic and social life, but also from basic health care.

To date, an estimated 800,000 people have been diagnosed as HIV positive out of a total population of 60 million. It is estimated that 30,000 people now die of AIDS every year, while 4,000 HIV-infected babies are born. Another 7,000 have sero-positive parents, thus risking the development of symptoms later on in life. Bangkok is only one of Asia's big cities where the problem has become highly visible. Apart from the obvious medical problems – the number of hospital beds occupied by AIDS patients is sharply growing – major social phenomena have emerged in Thailand with direct, negative effects on AIDS sufferers.

'AIDS will exact the heaviest toll from those who can afford it least: poor families who not only lose their main wage-earners, but also have to bear the cost of caring for them,' writes Gordon Fairclough, in the *Far Eastern Economic Review*, on 21 September 1995. In families with AIDS patients, wages are often swallowed up by treatment costs. It is therefore not unusual to find families rejecting members diagnosed as HIV positive. Very often it is the grandparents who have to take on the financial burden when parents are no longer able to work because of illness. They may also take on the care of their grandchildren, who might themselves by HIV positive. Indeed, it is believed that Thailand will have more than 95,000 orphans under the age of 15 by the turn of the century. These children will experience major difficulties in finding funds to finance any kind of higher education.

Families which cannot cope with the financial burden of income loss or treatment costs are at a high risk of being trapped in a hopeless cycle of marginalisation and exclusion. The situation is particularly uncertain for women. As one young widow in Chiang Mai noted: 'First our husbands get sick and we care for them until they die. Then we have to look after their parents and work to make ends meet. And when we are exhausted and sick, there is nobody left to look after us.'

It is not only the financial implications that push families to reject their sick relatives. There is also a fear that has been reinforced by aggressive campaigns using slogans such as AIDS = DEATH. These campaigns have stigmatised both the disease and the patients, forcing them into isolation and loneliness. Such ostracism represents one of the most terrible social and personal symptoms with which AIDS patients are forced to struggle.

MEDICAL AND PROGRAMME DISCRIMINATION

Patients who are diagnosed as HIV positive are regularly refused surgical interventions. No matter how minor the complaint, they are likely to be put into a ward where all the patients are either HIV positive or have full-blown AIDS or may even be in the terminal stages of the disease. In one example, a homeless person arrived in a university hospital following a bus accident. He had a swollen knee, oedema and scabies. An X-ray indicated tuberculosis and a blood sample was taken for analysis. It was discovered that he was HIV positive. He was refused admittance to the hospital because there were no beds available. He remained for three days in an observation room receiving only analgesics. No further TB tests or treatment were carried out and he was then transferred to a home for HIV-positive patients.

The fear of AIDS is so great that some communities oppose and physically intervene to prevent homes for AIDS patients being set up. The following example is far from an isolated case. A home for AIDS patients in Bangkok was being run by an NGO working closely with an AIDS clinic. At one point, the home came under pressure from the local community. The NGO coalition sought to counter the protests by running a health information campaign. At first, this appeared to be successful in lessening the tension, yet failed to prevent a grenade from being thrown into the home one night. A new campaign was started up. Once again, however, the home was attacked, this time with a machine-gun, and the car belonging to the project was destroyed. Although nobody was hurt, it was decided that the risks were too great and the home was closed.

Although similar phenomena appear elsewhere in southern Asia, countries such as Burma, Cambodia, Vietnam and China have tried to ignore the problem. They have also done less to disseminate information about AIDS prevention. As a result, they will have an even higher price to pay. In India, for example, the World Health Organisation already estimates at least 1.5 million HIV-positive cases. All will need to admit to the reality of AIDS spreading throughout their populations and find the means to provide at least a minimum level of medical and social care.

MSF decided to start its Thailand AIDS operation in 1993 as a home care programme for AIDS patients. This includes support for patients and their families as well as networking among hospitals, homes and families to find the most suitable solution for each patient. In charge of health education, treatment and consultations but also information on deontological malpractice, MSF operates in partnership with 47 other agencies of the Thai NGO AIDS coalition.

Stephan van Praet

to protect everyone. If our societies do not invest enough energy and money in the battle against exclusion, they will be forced to pay a very high price in the long run, as whole layers of their populations will slip into the twilight zone between citizenship and the marginalisation that forces them into criminal activities. Governments cannot risk counting only on the goodwill of private initiatives. They have to face their responsibilities and create a framework for a fair system of protection that prevents structural exclusion. If they do not want to do it out of respect for their citizens, they can do it out of self-interest, as criminal elements as well as political and religious extremists find fertile soil in socially excluded groups.

At the international level, the USA and Europe tighten their borders, refusing asylum and citizenship. At the individual level, we are building higher walls and more sophisticated security systems for our houses. These barriers can create an environment of indifference but they can neither exclude nor isolate us entirely from the world of which we are all a part. There is only one world, and there is only one type of citizen – unique.

This chapter was translated by Mrs Alison Marschner.

6

INTERNATIONAL LAW
AND REALITY
THE PROTECTION GAP

Iain Guest and Françoise Bouchet-Saulnier

•

THE CRISIS

One hundred and two Lebanese civilians die when Israeli troops shell a United Nations compound in southern Lebanon. Women and children are targeted by snipers in Bosnia. Civilians in Rwanda are forced to murder friends and neighbours as part of the genocide against the minority Tutsis. Such events, picked at random from recent newspaper headlines, are typical of the perils that face civilians in today's wars. The protection exists on paper and in legal texts, but in practice it is paper-thin. The charnel-houses are filled on four continents. Innocents are starved, expelled, humiliated and summarily executed. What exactly are we offering them – protection or betrayal?

This question should be asked as a matter of urgency after the wars in Bosnia and Rwanda. On the one hand, they have prompted a flurry of initiatives aimed at improving the protection of civilians. Two international criminal tribunals have been established to prosecute war criminals. The UN and Organisation of Security and Cooperation in Europe (OSCE) are deploying human rights field officers. With one eye on Bosnia, governments have drafted new international rules to curb the use of landmines.

UN agencies are trying to use their specialised mandates to mitigate the impact of war. UNICEF (UN International Children's Emergency Fund) commemorated its 50th anniversary in December 1995 by issuing a 12-point plan of action to give better protection to children in war. The World Health Organisation (WHO), the UN Development Programme, UNESCO (UN Educational, Scientific and Cultural Organisation) and the World Food Programme (WFP) are all looking at ways of responding more effectively to emergencies and wars. In spite of this activity, there is a sense of crisis at organisations like the office of the UN High Commissioner for Refugees (UNHCR) and the International Committee of the Red Cross (ICRC). To hear their senior officials speak on Bosnia or Rwanda is to conclude that humanitarian principles are under siege. Governments appear increasingly inclined to accept barbaric behaviour as the norm.

Part of this is due to a change in the nature of warfare. Today's wars take place within, rather than between, states. Only 17 per cent of the casualties in the First World War were civilians: in today's wars, the figure is over 90 per cent. The nature of weaponry has also changed, increasing the risk to civilians. An estimated 110 million landmines have been laid in over 50 developing countries, with little regard for their impact on civilians. This impact, as a result, is devastating: 26,000 new amputees a year. In addition to targeting civilians during a conflict, landmines extend the threat way beyond the peace agreement.

But the fundamental crisis facing civilians today comes from the attitude of combatants, who display a contempt for the rules of war. The philosophy of war has evolved. Today, we see less and less direct confrontation between combatants, as armies aim for zero casualties. But saving the soldier may condemn the civilian, because the objective of war today is less to win than to exterminate the enemy and wipe his territory off the map. Civilians bear the brunt. Sniping at children, raping women, and torturing prisoners are outlawed by the Geneva Conventions, but to many of today's warlords they are accepted military strategy. Not for nothing has 'ethnic cleansing' become the phrase of the 1990s: after all, it refers to the mass expulsion of civilians. Targeting civilians has replaced torture as the main instrument of state terror.

How should the humanitarian agencies react? Some insist that assistance in times of crisis must be kept separate and distinct from protection. Whatever their intention, and however courageous the actions of the relief organisations in the field, this relieves governments of the obligation to speak out against the barbarism, and so ends up by sanctioning it.

AID AS PROTECTION

Humanitarian relief agencies are usually the first to arrive on the frontline, bringing with them an important form of protection in the shape of their emergency relief supplies. Governments accepted this role by agreeing, in 1949, not to consider impartial aid as interference in their internal affairs. Despite this, many relief agencies prefer to concentrate on the technical challenges of delivering relief aid, rather than protection. Only two aid agencies – Médecins Sans Frontières and the ICRC – were present in Rwanda in 1994 during the genocide. Yet months later, when the protection crisis had passed, their number had grown to over 140. After virtually ignoring the genocide in Rwanda, relief agencies have spent well over a billion dollars on its aftermath.

Under pressure from donors to show that relief aid is being well spent, non-governmental and UN aid agencies are preoccupied with improving the delivery and coordination of relief aid. Some deal with water, others with vaccinations, others with the distribution of medicine, food and seed.

Some care specifically for children, others for refugees, still others for prisoners. But this specialisation relieves them collectively of responsibility for the fate of the victims.

UN organisations are even more likely to lose sight of the real priority, because they are inter-governmental organisations that have been created to promote technical cooperation with governments. They depend upon the goodwill and cooperation of national authorities to be able to act in emergencies. When a national authority is bent on the total destruction of an opposition, or ethnic group, there is not a lot of goodwill around.

THE MANIPULATION OF AID

In spite of their obvious limitations, relief agencies are increasingly called upon to assume the task of protecting civilians. For the first three years of the war in Bosnia, none of the major powers was prepared to use force to stop ethnic cleansing by the Bosnian Serbs. Instead, they assumed that the relief agencies, led by the UNHCR, would contain the violence and prevent a mass exodus of refugees, merely by dint of delivering emergency relief aid. This was known as preventive protection.

Neither UNHCR, nor its lightly armed escorts from the UN Protection Force (UNPROFOR), were up to the task. Often, UNHCR officials would be forced to make a tough choice between delivering food aid or protesting abuses. Faced by the possibility of famine, and unable to call in military support, there was really no choice: food aid won out. Ethnic cleansing continued for over three bloody years, culminating in the massacres at Srebrenica in the summer of 1995.

Humanitarian aid has been manipulated also in the case of Rwanda. Despite the presence of a UN observation and assistance mission in Rwanda, as many as a million Rwandans, mainly Tutsis, were murdered in 1994. Those responsible fled into neighbouring countries, together with 2 million Hutus. Here, they have continued where they left off in Rwanda, confiscating aid and intimidating refugees.

UNHCR officials have had no option but to rely on the local authorities in the countries of first asylum. In the case of Zaire, this has meant soldiers from President Mobutu's Presidential Guard, who have been hired by UNHCR to provide security. The results have been meagre: a handful of intimidators have been seized and removed from the camps. But hundreds of former killers (*génocidaires*) still live openly and exercise control over camp committees. The situation in these camps is more stable than in 1994, when murder and thuggery were commonplace; but it is the balance of terror that has stabilised, rather than the protection of innocent refugees.

FAILED EXPERIMENTS

In order to hide their unwillingness to protect civilians, and their total lack of strategy, governments have resorted to launching spectacular relief operations in recent years. Their aim, however, has been more to assuage public opinion than to protect. This has led to several experiments which were presented as protection solutions but in fact were acts of desperation. In the former Yugoslavia, the UN Security Council declared several 'safe areas' but then found that it could not find the troops or the political will to protect them. The Bosnian Serbs took their cue and turned the 'safe areas' into hellholes: Zepa, Gorazde, Sarajevo, Bihac, Srebrenica.

Similarly, in June 1994 the UN Security Council accepted a French proposition to create a secure humanitarian zone inside Rwanda (Zone Turquoise). French troops were sent in to protect the civilians and discourage any exodus towards Zaire. But the intent was less to provide sustained protection than mount a dramatic, headline-catching intervention on behalf of France's clients – the Rwandan Hutu – who were now losing the war. The departure of the French resulted in a predictable tragedy, when the Rwandan army dismantled the camp of Kibeho. Several thousand displaced Hutu were killed in a chaotic, bloody stampede. Three months later, in Bosnia, UN soldiers abandoned the 'safe area' of Srebrenica that they had pledged to protect, and thousands of Muslim men and boys paid with their lives. It was a ghastly failure.

DILEMMA FOR THE AGENCIES

But while Bosnia and Rwanda show that relief aid on its own cannot provide protection, they also present the aid agencies with some difficult choices. Is there any way they can continue to perform their task while avoiding manipulation? Do they have any room for manoeuvre? Any room for free expression? To some extent, all non-governmental aid agencies face this dilemma. MSF, for example, took a risk in Rwanda in 1994, by denouncing the genocide while still maintaining a field operation in Kigali. The ICRC has traditionally relied on discretion and neutrality when operating in difficult and sensitive war zones. Usually this is a precondition for being allowed to operate. But Bosnia has also shown that both can be exploited by unscrupulous combatants, who try and use Red Cross visits to give an impression of normalcy. As a result, one sees the Red Cross becoming more vocal in denouncing violations of humanitarian law, and bolder in taking up controversial issues like the campaign to ban landmines. But it remains to be seen how far the ICRC can go without provoking governments. This is more likely if agencies can be played off against one another by governments. The ICRC has such a specific mandate under the Geneva Conventions that it cannot be replaced by another organisation. But the ICRC is unique in this sense. There are no common guidelines among aid agencies and, as their numbers multiply, so it becomes easier for governments to play off one against another – so that those organisations that openly criticise government authorities are liable to be harassed to the point at which they can no longer work. It is no coincidence that when governments are trying to rationalise aid, they often expel the most outspoken agencies and tolerate the most docile. This was the case in December 1995, when the Rwandan government expelled 38 NGOs in an effort to quash NGO criticism of its actions. This presents the agencies with yet another dilemma: it is often the case that, when the authorities feel the need to expel aid agencies, a population is suffering more from terror than a shortage of food and other basics. This may well be another reason for withdrawing.

How can the agencies respond? One way might be by putting more emphasis on protection and freedom of expression. MSF has shown time and again that it is possible to speak out against abuse and also maintain a field presence. This was the case in the former Yugoslavia, and also in Rwanda. At the same time, MSF does not view relief as an end in itself, particularly as it can easily be manipulated. This is why the organisation decided to withdraw from providing relief in the refugee camps of Zaire rather than see the relief used by those who had perpetrated the genocide.

Once again, the problem is more acute for UN organisations because of their inter-governmental nature. Even when they possess a clear humanitarian mandate, as in the case of UNHCR or UNICEF, they cannot oppose governments or impose their own will. The same is true of military efforts on behalf of humanitarian operations. The UNPROFOR in the former Yugoslavia emerged from Bosnia with its reputation in tatters because it gave the impression of siding with the Serb aggressors against the Muslims, in an effort to preserve neutrality.

But once peace-keepers try to impose their will by force, in a situation where there is no central government, it can lead to the opposite kind of problems, and terrible suffering for civilians. This, surely, is one of the lessons from the second UN force in Somalia (UNOSOM 2), which killed scores of Somalis in an effort to track down the warlord Mohamed Aideed.

One thing is certain. On their own, relief agencies cannot and must not assume that their mere presence will protect civilians in conflict. Indeed, Bosnia and Rwanda suggest that by even attempting to do so they may increase the risk. But whether this means they should withdraw, rather than submit to manipulation, is another matter altogether. Several NGOs withdrew from the Rwandan refugee camps in protest against feeding *géno-cidaires*. But UNHCR decided to stay, on the argument that withdrawal would only have exposed the weak and vulnerable to starvation. To pull out is the ultimate sanction, and it is far from certain whether it will save lives. If the warlords and guerrilla leaders call the bluff, then the answer is probably no.

There are, in short, no magic solutions, and no escape in neutrality or withdrawal. In spite of this, aid agencies should, at the very least, acknowledge the potential for manipulation, and try and stake out parameters for their involvement.

HUMANITARIAN LAW

Humanitarian law has provided the legal basis for 'humanising' war for well over a hundred years. On paper, it has attained near universality and is moving into exciting new areas. In practice it is not expanding nearly fast enough. Humanitarian law was born after a young Swiss businessman, Henri Dunant, happened upon the aftermath of a major battle in northern Italy, Solferino. Appalled by the piles of dead and dying, Dunant persuaded the Swiss government to host a meeting of governments to discuss ways of protecting wounded soldiers. Twelve attended in 1864, and agreed to treat each other's wounded in battle, thus giving birth to the first Geneva Convention. An organisation was formed and a new emblem found for it by turning the Swiss flag – a white cross on a red background – inside out.

This was a pragmatic start for the Geneva Conventions and the Red Cross movement, and they have remained wedded to pragmatism ever since. The goal of the Conventions is to protect non-combatants in war, but this is done by appealing to self-interest – not by invoking inalienable rights or natural law, as is the case with the human rights approach. The 12 governments agreed to protect their enemy's wounded in 1864, precisely because their own wounded might need protection.

The codification of humanitarian law was also helped by the evolution of democracy. Instead of being viewed as cannon fodder, soldiers came to be seen as citizens who would be badly needed back home once the war finished. This self-interest encouraged governments to draft two subsequent Conventions, on shipwrecked sailors and prisoners of war.

There is another aspect to this pragmatism. Humanitarian law attempts to mitigate the damaging consequences of war, not put an end to the war itself. Nothing in the Geneva Conventions challenges the time-honoured rationale for war – that it may be needed to settle disputes. The aim is to limit the damage. Confrontation between troops is thus permitted, but at certain times, in certain places, and according to the rules of the Geneva Conventions.

By forsaking a higher moral standard in favour of self-interest and pragmatism, the Geneva Conventions have been able to attract nearly universal membership. But self-interest is no longer enough if one side is sufficiently cynical, sufficiently indifferent to international outrage, or feels itself sufficiently strong to impose its will without fear of reprisals – as was the case with the Bosnian Serbs between April 1992 and July 1995. Often, today's warlords act as if they have nothing to lose by breaking the laws of war, and nothing to gain by respecting them.

PROTECTING CIVILIANS

By the time of the outbreak of the Second World War, the Geneva Conventions protected three important categories of non-combatant: wounded soldiers; prisoners of war; and shipwrecked sailors. The glaring absence was civilians, but when the ICRC proposed to draft a new convention covering civilians in 1938 it was vetoed by governments on the grounds that it would limit their military options. The Germans in particular took full advantage of this loophole in the war that followed.

The cruelties of the Second World War prompted governments to draft a new Convention, in 1949, extending protection to civilians in war zones. This was an exciting development, because up to this point a country's civilians had been considered an internal affair. Now, their protection would be a responsibility for the entire international community. The Conventions spell out, in detail, those guarantees to which civilian populations are entitled in times of war. Moreover, it specifically describes the organisation of aid in such situations. But the 1949 Geneva Conventions only apply in wars between states. They do not apply in wars within states, be they ethnic, civil, or wars of liberation. The Vietnam war showed that something else was needed, and between 1974 and 1977 a plenipotentiary conference of governments drafted two Additional Protocols which expanded coverage of the Conventions to internal armed conflicts and also expanded the protection measures afforded to non-combatants. The problem is that adherence to the two Additional Protocols has grown slowly. Several major governments, like the United States, have yet to ratify. In addition, although the protocols set a threshold of military activity above which a conflict qualifies as international, it is left to the government to decide when this point is reached. Given the choice, most governments will not invite international scrutiny of their behaviour, in case this limits their fighting ability. This is why the government of the Russian Federation is able to declare the current warfare in Chechnya as an internal affair, and why this qualification alone relieves the Russians from applying humanitarian law. The result is mayhem. In Chechnya, the civilian population is accused of supporting the fighters. On this pretext, whole villages are pillaged and razed to the ground, and aid organisations are forbidden to enter bombed villages to pick up the wounded.

For a brief period in the early 1990s, it appeared as though the doctrine of national sovereignty was cracking and that this might open the way to more humanitarian intervention. The French government spearheaded an attempt to get the UN General Assembly to accept that endangered civilian populations had a right to humanitarian assistance, but Third World governments were able to blunt this initiative, and ensure that the donors were not given a licence to deliver aid to their foes.

Sovereignty was weakened, however, in 1991, when the Western Allies delivered humanitarian aid to the Iraqi Kurds in northern Iraq against the will of the Iraqi government. Security Council Resolution 688 (5 April 1991) authorised the allies to deliver humanitarian aid to the Iraqi Kurds in northern Iraq with or without the permission of the Iraqi government, but this would have meant little without the presence on the ground of the Coalition force that had just won the Gulf War. The Soviet Union and Yugoslavia then fell apart. Everywhere, devolution seemed to be the order of the day. In June 1992, the UN Secretary-General further chipped away at national sovereignty by concluding in his Agenda for Peace: 'The time of absolute sovereignty has passed: its importance was overstated.'

Would this make it easier to reach beleaguered civilians? For a time it seemed so, as UN peacekeepers were sent into a series of countries (Cambodia, Somalia and Bosnia) which had no central government to speak of. Their mandate was to monitor cease-fires, deliver aid to civilians caught up in conflict, and try to build peace. Looking back, none of these three are particularly successful examples of international intervention. The fact that Russia can successfully insist on keeping the war in Chechnya an internal affair suggests that the doctrine of state sovereignty may be making a comeback.

LANDMINE DISAPPOINTMENT

Several important initiatives have been launched in an effort to improve the protection of civilians. To take one example: governments met in Geneva in March 1996 to revise a 1980 protocol limiting the use of landmines. The new text appears to tighten up the rules, by calling on governments to ensure that all anti-personnel landmines are detectable and that all remotely delivered mines are built to self-destruct or self-deactivate within four months.

This might seem an improvement over the current anarchy. The problem is that it leaves the mapping, marking and removal of mines to those who laid them – which means vicious war leaders like Ratko Mladic, the Bosnian Serb commander. Even more important, the new protocol allows the production of smart mines which self-destruct. This undercut the courageous decision of almost 40 governments that have agreed to ban all production and use of mines. Nor will it necessarily reduce the risk to civilians, because modern mines systems are capable of delivering thousands of smart mines in minutes. To a Chechen farmer whose valley is being carpeted by mines, it makes little difference whether they will self-destruct within four months. If they don't the new protocol makes no provision for calling those responsible to account, because it contains no provision for enforcement.

Ultimately, this is why humanitarian law is under siege. Instead of confronting today's warlords, the great powers are trying to co-opt them. It will not work. They cannot be trusted. What can be done? Much more use could be made of public opinion. Law – any law – is not a body of texts. It is a living legal instrument that has to be argued. In this, public opinion and the media can play an important role in helping to categorise situations of conflict correctly, and highlight the needs of victims. Humanitarian agencies are not nearly as aware of this as they should be. Indeed, many know little about humanitarian law.

In addition, the Geneva Conventions provide several avenues which could be more widely used, or widely understood. The application of humanitarian law by a state in conflict does not imply the legal recognition of its adversary. Regardless of the nature of the conflict, and regardless of whether the adversaries have signed the Geneva Conventions, they may sign a special accord in the course of the conflict to guarantee the application of humanitarian law protecting civilian populations. This allowed the ICRC to obtain a written agreement from the Serbs providing for the protection of civilians in the former Yugoslavia, even though Belgrade considers the conflict in Bosnia to be an internal affair.

In the event, however, it is doubtful whether this agreement had any more practical impact than all the other pledges and cease-fires offered by the belligerents in this brutal war. Why? Because combatants in today's wars feel increasingly less inclined to fight by the rules of war. How can they be brought back on course?

HUMAN RIGHTS – SETTING THE STANDARDS

While humanitarian law is essentially pragmatic and based on self-interest, human rights are inalienable and universal. In legal terms, the difference is crucial. The Geneva Conventions do not proscribe the killing of one soldier by another, but rather lay out how it shall be done. But the right to life, as laid out in human rights instruments, is sacrosanct.

Human rights pose a direct challenge to governments by defining and prohibiting governmental abuses against individuals and groups. Governments are asked to protect rights by ratifying treaties. Once they agree, they can then be held accountable for violations. There is one important exception, in that states can derogate from some of the provisions in times of emergency. But no derogation is allowed for core rights, such as the right to life. (This is similar to humanitarian law, in that Common Article Three of the four Geneva Conventions applies at all times, regardless of the categorisation of the conflict in question.)

In theory, these human rights standards should be a powerful weapon in the fight to protect civilians against governmental forces, particularly given the nature of contemporary warfare. The wars in Bosnia and Rwanda were not the product of 'blind ethnic hatred', as they are sometimes portrayed. They were fanned and exploited by two ruthless regimes: the Hutu-dominated regime of President Habyarimana in Rwanda, and the government of President Slobodan Milosevic in Serbia. As the 1995 annual report of Human Rights argues: without stimulation by opportunistic governmental leaders, communal tensions rarely rise to large-scale violence.

The horrors of the Second World War produced a burst of international legislative activity that resulted in the Universal Declaration of Human Rights (1948), the Genocide Convention (1948), the 1951 Refugee Convention and the two international covenants (on civil and political, and economic and social rights).

The UN has constructed a formidable edifice of committees, rapporteurs and working groups to monitor the application of these and other standards. In theory at least, many could be used in the protection of civilians in war. But once again, as with humanitarian law, the loopholes and shortcomings are glaringly obvious.

GENOCIDE

In 1948 governments adopted a convention outlawing the crime of genocide in the hope of protecting civilians not only from the effects of war but from acts of extermination as well. The definition of genocide did not include political groups, and this has limited its application. Nor – unlike the Geneva Conventions – was any specific means of application envisaged. No means of aid to victims is specified (since it is impossible to assist victims of genocide). The convention simply states that states are individually and collectively committed to put a stop to acts of genocide.

The weakness of this approach was demonstrated in the massacres which erupted in Rwanda in April 1994. For several months, the international community simply refused to qualify a massacre as an act of genocide, thereby avoiding the obligation to stop it. Four months after the killing stopped, the UN Security Council finally got round to qualifying the situation as an act of genocide. States then promised to punish the perpetrators of genocide. It is a strange justice, in which witnesses at the inquest are all dead.

REFUGEES

The 1951 Refugee Convention provides for the right to seek asylum. The UNHCR assists 19 million refugees today, 4 million of whom fled their countries in 1994 – and of these, 2 million were Rwandan. Would-be refugees have found it increasingly hard to win asylum since the end of the Cold War. After an initial exodus from Croatia and Bosnia, it proved almost impossible for victims of the Bosnian war to find sanctuary outside the arena. Why this hostility towards asylum-seekers? It is not that the world is a safer place. Certainly, it may be harder for asylum-seekers to demonstrate an individual fear of persecution (as required by the 1951 Convention) than it was during the Cold War. But this has been fanned by right-wing extremists, and exploited by governments who pander to anti-immigrant sentiment.

As a result, there is increasing pressure on asylum-seekers to return to their country of origin as soon as possible, which may be sooner than is safe. Once refugees return home, they forfeit their refugee status, which means they have to rely on the same degree of protection as is available to ordinary citizens. Although millions have returned home safely since the end of the Cold War, they have not found it easy to reintegrate. Bosnians and Rwandans are returning to ruined homes, suspicion and even mines. At least on paper, asylum is still a recognised international human right. An even more serious gap in the protection of refugees stems from the fact that internal armed conflicts generate internal refugees who are not covered by the 1951 Convention because they never leave their country. According to UNHCR, there are 27 million of these internally displaced. While they clearly have special protection needs, they are not recognised as such under international human rights law. Nor are they even assured of consistent treatment by the international system.

UNHCR assumed responsibility for feeding internal refugees in Bosnia, but has declined to assume the task in Burundi, fearing that it would be overstretched and sucked into a military and political black hole. A special rapporteur of the UN Secretary-General, Francis Deng, monitors the plight of IDPs (internally displaced persons) for the UN Human Rights Commission each year. But Mr Deng's power is limited to his reports, and they do not lead to intervention by the Security Council. It is yet another illustration of the protection gap.

CHILDREN

The UN Convention on the Rights of the Child outlaws many of the abuses practised against children in war and calls for the special protection of vulnerable categories such as refugee children and migrants. The Convention has more ratifications than any other human rights treaty, and UNICEF is using it to campaign against the recruitment of child soldiers and even the indiscriminate use of landmines. But the appalling cruelty meted out to children in Rwanda and Bosnia suggests that describing their murder as a violation of the right to life will have no more practical effect on combatants than pleading for the Geneva Conventions to be respected.

HUMAN RIGHTS IN ACTION

The first and most notable feature of the UN's human rights bodies is their detachment. Signatories to treaties submit regular reports, which are then examined in committees. Individual governments that practise a systematic pattern of gross violations are singled out for discussion by the UN Commission on Human Rights that meets every year in Geneva. Although the Commission has deployed an expanding number of rapporteurs such as Francis Deng, who conduct on-site investigations and produce valuable reports, it is far removed from the killing fields of Rwanda, Burundi or Bosnia.

Since the end of the Cold War, this archaic approach has been complemented by the deployment of human rights field monitoring teams. The first such mission, ONUSAL, was sent to El Salvador in 1991 before there was even a cease-fire between the government and FMLN (Front Farabundo Marti de Liberation Nationale). The second worked in Cambodia in 1992 and 1993.

Subsequent missions have been deployed in Haiti and Rwanda. These were not humanitarian operations. The human missions in Haiti, Cambodia and El Salvador were all sent by the UN Security Council under its peace-keeping mandate or as part of peace-keeping missions. This enabled them to draw on an extensive range of assets, including money, military observers and UN civilian police. With the exception of Rwanda, they also had a finite goal: to monitor and improve human rights conditions in the run-up to elections. The success of these missions varied. It is often the lack of such fundamental basics as notebooks, door handles, registers and type-writers which contribute to the abuse and torment of prisoners in overcrowded jails. (The lack of handles to shut doors, for example, means that prisoners have to be shackled.) Often, these problems may be better addressed by a sympathetic field monitor with a little petty cash than an elegant resolution at the UN Human Rights Commission, thousands of miles away in Geneva

But these monitors also face the same dilemma that confronts aid agencies: in order to operate, and gain access to jails, they may have to be discreet and respect confidentiality. This weakens their leverage with the authorities responsible for violations. All they can do is request an end to the abuse. Not only is it hard to strike the right kind of balance, but it will depend on factors beyond the monitors' control. The UN mission in Cambodia was able to persuade the State of Cambodia to remove shackles and release scores of prisoners from jails in its first few months in Cambodia in 1992. But the SOC authorities then grew resentful that the Khmer Rouge were not cooperating – and hence not submitting to criticism. By the time the monitors withdrew, none of the Cambodian groups were whole-heartedly working with them.

Discouraged by the limited results and high cost of recent UN peace-keeping missions, governments are less inclined to deploy Cambodia-style peace-keeping operations. This means that the burden of human rights field monitoring will be assumed by the Geneva-based UN Centre for Human Rights. The Centre sent a small team to the former Yugoslavia during the war, but it is in Rwanda that it has launched the largest and most ambitious experiment.

Plate 6.1
Due process remains non-existent in this prison in Kigali, Rwanda. Photograph: Andreas Herzau

EXPERIMENT IN RWANDA

At the time of writing, 101 human rights monitors are working in Rwanda. This is almost 50 below the target, but still considerable for a UN Centre that has been chronically mismanaged and underfunded for the last 20 years. This operation was a personal initiative of José Ayala Lasso, who was appointed to the newly created post of UN High Commissioner for Human Rights just before the genocide erupted in Rwanda. The mission suffered from all the problems associated with a first experiment. It took time to recruit, time to sort out administrative wrinkles between Geneva, New York and Kigali.

At the time of writing, the field mission in Rwanda has consolidated. Its members have acquired valuable on-the-job training and built up solid

Unending hate

I witnessed the massacre at Kibeho, and I can still see in front of me frightened faces and expressions of panic. Whenever I close my eyes, I see scenes I want to forget. I can still hear the cries, the blows, the gunshots and the machetes. I still remember running under the gunfire and seeing my friend trapped in a doorway where I could not go to her. But there is worse than the fear, the carnage and the horror that I saw – that is the conclusion that it will never end.

These people are so full of hate that they will kill each other at the slightest provocation, even with so many onlookers, whatever we do or don't do. Sometimes it will be the Hutus and sometimes it will be the Tutsis. They will switch back and forth between tormentor and victim, but they won't stop. They don't want to stop. They have no motive other than revenge, and this revenge has become a never-ending spiral. This is what I feel like after five days of hell. In fact I don't just feel it, I *know* it. It *will* happen again. And I feel so sorry that nobody has the means, the will or the ability to stop it. It is hard to return home after such an experience, to get back into the routine of work and maintain my motivation. It's impossible to forget, although I know that to go on I have to rid my mind of the pictures and feelings I built up over so short a time. What I am left with is the terrible frustration of being powerless to change the slightest detail of this story of unending hate, one which we will never be able to understand.

Laura Barroeta, Spanish doctor,
MSF Kibeho Project, Rwanda

relationships with local leaders. This could enable them to monitor the return of the Hutu refugees, if and when this occurs. But there are also two huge question marks hanging over their work. The first comes from funding. The mission depends entirely upon voluntary funding, and has only been able to attract funds for one or two months at a time. No monitor has a contract for longer than a month, making it impossible to ensure any continuity.

The other, more profound concern stems from the mission's mandate. Is it there to observe or to protect? This is a perennial concern. Many Haitians came forward to report violations to the UN/OAS (Organisation of American States) mission in 1993, at considerable personal risk. They were left vulnerable and exposed when the mission was forced to withdraw in October. The same happened in the early days of the Rwandan mission. Rwandans who took considerable risks to report delicate information received no support from these defenders of human rights. Those who took refuge in the locale of the UN human rights observers' mission were turned over to Rwandan police.

The key to this dilemma lies in the relationship between the monitors and the Rwandan authorities. On the one hand, they have struck up a dialogue with the authorities in Kigali. On the other hand, they openly report abuses by the authorities: the massacre of over 100 civilians at a commune called Kanama; the arrest of the deputy prosecutor of Kigali on a vague charge of genocide; the unacceptable conditions in Rwanda's overcrowded jails.

This is an innovative diplomatic high-wire act. But it is not protection. There is little evidence of any improvement in Rwanda and some very worrying reports are coming out. In the last two months scores of desperate prisoners have died from suffocation, overcrowded conditions or while trying to break out of jails. The justice system remains paralysed. Returning refugees are being arrested.

Perhaps the task is beyond the capacity of 100 human rights monitors. It is certainly not easy to heal the wounds of mass genocide caused by a past regime, while at the same time preventing abuses by the current government. The problem is, that by their very presence, the monitors have pledged to try.

Plate 6.2
Chain gang of (criminal) prisoners after taking a bath in the Irrawaddy River near Myitkyina in Burma's Kachin State. Photograph: Jan Banning

AID AND HUMAN RIGHTS

Looking at Rwanda, one has to conclude that by being deployed in the field, the monitors have made the UN seem less remote and detached. But they have yet to produce a dramatic improvement in the protection of ordinary Rwandans against abuse. They are also forced to make the same kind of compromises as the relief agencies in order to operate.

This situation may help to set priorities. One cannot consider all rights on the same scale. How many restrictions should one accept, and to defend whom, or what? As with relief aid, political, financial, military and operational constraints create their own priorities. Once again, in the case of UN organisations, these may not improve the protection of populations in danger.

The result can be dangerously misleading: while public opinion may be reassured by the presence of human rights observers in the field, the victims see these observers as a smokescreen whose real aim is to protect the security of the country rather than the people. Politicians seem to be setting the priorities, which the jurists then defend.

THE CRISIS OF ENFORCEMENT.

Human rights and humanitarian law can establish standards, but who is to transform them into protection? Without police, judge and jury, laws mean nothing, and this applies equally to the two international human rights Covenants and the four Geneva Conventions. These laws will not protect anyone if no one is prepared to enforce them. In some desperation, we resort to rhetoric and invent new phrases and definitions, as if calling an

---

We always collect firewood around our village. One day, Laila, my 8-year-old sister, and I went outside the village to look for firewood. We found an area with a lot of wood and that made us very happy. We rushed out to collect all the wood that was there.

After a few minutes Laila shouted, 'Hey, I found a beautiful toy! It's my toy and I don't want to share it with you.' I protested and said, 'We share the wood so we should share everything we find.' 'No, no,' she said, but I ran and caught her. She tried to pick up the toy. At that moment I heard a loud sound and fell on the ground. I saw blood on my hand. Two of my fingers were gone. I rushed to see Laila, though I was in a lot of pain. She had her hands over her eyes and was crying. She had lost both her eyes.

It was a butterfly mine, a killing toy.

Naiza, ten years old, Gozara district,
Herat Province, Afghanistan

emergency 'complex' or 'integrated' will somehow make it easier to deal with. Human rights and humanitarian experts ask questions but find few answers, and each finds in the other what he lacks. The human rights community sees a practical ability to operate in the field, while relief agencies envy the ability of the human rights community to denounce violations.

But is this progress or mutual reassurance? Perhaps it would be better if both saw it from the perspective of the victims. Otherwise, it becomes another excuse not to act: genocide in Rwanda is conveniently redefined as a complex humanitarian crisis and the most barbarous attacks on civilians in Bosnia are tolerated by the UN in the interests of preserving world peace. Similarly, focusing on the coordination of aid can result in protection being pushed to the side. Coordination dilutes the sense of responsibility any participant may feel with regard to the victims.

Instead of trying to facilitate cooperation and coordination between international organisations, we should reaffirm and reassert the fundamental principle that populations in danger require protection. This, however, will mean confronting the contempt and indifference shown towards law by today's combatants. If one shortcoming stands out, it is the lack of enforcement. This weakness is common to human rights and humanitarian law. It has weakened the 1948 Genocide Convention, the 1949 Geneva Conventions, the 1991 Convention on the Rights of the Child, and virtually all other human rights treaties. The Geneva Conventions, for instance, call for a system known as the 'protective power,' under which states not involved in the conflict accept the responsibility to defend the interests of the civilian population in accordance with the norms outlined in the fourth convention. This is pragmatic enough, but since the adoption of the 1949 Conventions it has never been put into practice – one more casualty of the doctrine of state sovereignty and non-interference. Even when this interference is legitimate, authorised, and necessary, no one will volunteer. Usually, the ICRC is left with the task of acting as protective power – a role that it cannot possibly fulfil, given its limited resources and carefully defined mandate.

The Geneva Conventions also stipulate that the authors of serious violations will be sought out by all member states, and that they could be judged by any court in any country. Despite this commitment, no state has yet prosecuted a criminal from a foreign war. The same lack of enforcement cripples human rights law. Violations of the 1951 Refugee Convention are reviewed on an informal basis by the UNHCR's executive committee, an advisory body which comprises many of those governments that are busily toughening curbs against asylum-seekers.

THE STRUGGLE AGAINST IMPUNITY

The most significant contemporary development in humanitarian law and human rights law has been the establishment of two tribunals to prosecute

Displaced populations: a vulnerable group in search of protection

The number of displaced persons has increased dramatically over the last five years to an estimated 25 to 30 million. Many have fled conflict areas because their most fundamental human rights have been put at risk. It is widely recognised that internally displaced persons are among the most vulnerable groups. They are regularly deprived of protection and vital services such as food aid and medical assistance and are often subjected once again to human rights violations similar to those which caused their displacement. Displaced persons, as distinct from refugees, remain within their country's borders. Their own governments therefore bear primary responsibility for meeting their protection needs. Yet displaced persons often fail to get adequate protection and assistance, sometimes because their governments are unable to help, but more often because they are unwilling to do so.

While refugees, having crossed a border, find institutional protection and material assistance through the UNHCR, among others, as well as legal protection in refugee law, no comparable system is in place for the internally displaced. The Department of Humanitarian Affairs and the UN Development Programme have been assigned to coordinate the provision of humanitarian assistance to the displaced, but neither has the mandate to provide protection.

A tragic example of the consequences of having no institution with a mandate to protect the displaced was seen in Kibeho, where thousands of the internally displaced were massacred by the Rwandan army in full view of representatives of several UN agencies, including UNAMIR soldiers. UNAMIR, the UN Assistance Mission to Rwanda, had a suitable mandate, but failed to take appropriate military action and was prevented from even speaking out about the killings. While human rights and humanitarian law contain numerous provisions that give protection to the internally displaced, international law has failed to establish mechanisms to protect those rights. Human rights law applies only to states and is not binding on non-state actors such as insurgents. And when an emergency threatens the integrity of the state, the state is not always bound to follow the provisions of human rights law.

The internally displaced depend on access to international assistance. While the internally displaced have, according to international law, the right to request and receive assistance, their governments are not obliged to accept such offers of assistance by humanitarian organisations or to allow free passage for humanitarian relief.

A clear example here is the Sudanese National Islamic Front's actions with respect to the internally displaced living around Khartoum. NGOs have repeatedly offered to provide humanitarian assistance to the displaced camps, and some projects have indeed been implemented. Yet the government has blocked badly needed expansions of these programmes.

> The government has denied access for proper needs assessments, has delayed the approval of project proposals and has not granted permission for an adequate number of expatriate staff to run the programmes. Medical data have established a clear link between the government policy and the alarming health conditions of the displaced.
>
> A legal provision establishing the duty of such governments to accept offers of assistance by humanitarian organisations and to grant and facilitate free passage of relief is therefore urgently needed. Providing the displaced with food, medical assistance and shelter on an *ad hoc* basis is no longer enough. Institutional, legal and on-the-ground protection is not an option but a necessity.
>
> Hanna Nolan, Dutch Policy Adviser, Humanitarian Affairs, MSF

war criminals from the former Yugoslavia (24 May 1993) and Rwanda (8 November 1994). These two bodies address some gaping holes in humanitarian law and the protection of civilians. In the first place, they take legal notions like 'genocide' out of the realm of theory and into the law court. Genocide was defined by a 1948 Convention, but the lack of a criminal court meant that it could not be implemented. Both of the tribunals have a mandate to prosecute practitioners of genocide in Bosnia and Rwanda, and both have invoked genocide in indictments. This is a functioning definition of genocide that applies to contemporary ethnic conflicts instead of the unique campaign conceived against the Jews by the Nazi leaders in the Second World War.

In the second place, the two tribunals advance the notion that individuals are responsible for their actions and cannot plead that they are acting under orders. Third, the tribunals provide the international community with a way of punishing war crimes (defined as grave breaches of the Geneva Conventions).

The statute of the Rwanda tribunal is also innovative in that by covering the massacres in Rwanda it is explicitly covering an internal armed conflict. The question is, will the tribunals seize the opportunity? There was tremendous scepticism when the first tribunal was established, during the height of the Bosnian war, because it was imposed on the General Assembly by the UN Security Council. This raised concern that it would be a substitute for tough action by the Council to stop ethnic cleansing.

Up to June 1996, the tribunal has indicted 57 individuals. Seven are in custody in The Hague or elsewhere. One trial is under way. The Rwandan tribunal is less far advanced: it has issued 11 indictments, but the deputy prosecutor in Kigali faces an uphill task in launching credible prosecutions: he has less than a third of the investigators he needs, and insufficient numbers of cars, interpreters and translators. There are many reasons to be

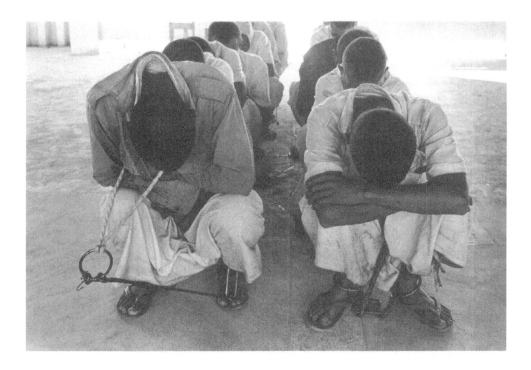

Plate 6.3
Prisoners wait to
start work as forced
labour in the building
of an army-run
market hall in
Myitkyina, Kachin
State, Burma.
Photograph: Jan
Banning

sceptical of these tribunals. They were established by the UN Security Council under its mandate for international peace and security, and could just as equally be disbanded by the Security Council which set them up, if they are deemed an inconvenience to peace. Their ability to investigate depends totally on the willingness of governments to cooperate; at some point this may collide with the fact that genocide and ethnic cleansing are launched by governments, not individuals. Investigations may also uncover links between the authors of state terrorism in Rwanda and the Balkans and other supposedly neutral governments.

The two tribunals will cost, together, $80 million this year – five times as much as the underfunded UN human rights programme. At the same time, they seem to lack basic resources. It will be difficult, or even impossible, to ensure the protection of witnesses and victims who are called to give evidence, particularly those from the former Yugoslavia. (Indeed, one rape victim has already changed her mind about giving testimony, out of fear of reprisals.) The rules of the Hague Tribunal lean over backwards to ensure that the rights of the defendants are protected – to such an extent that it may be hard to win the case against these suspected killers.

All this inclines some NGOs to keep their distance from the two tribunals. But the Hague Tribunal has also achieved something very important by labelling the Bosnian Serb leaders as war criminals, thus ensuring they are seen as international outlaws. In the case of Rwanda, the governments of Zambia, Belgium and Cameroon have all cooperated to arrest some notorious Hutu *génocidaires*. They include Theoneste Bagosora, who was chief of staff for the Rwandan Defence Ministry in April 1994 and as such presided over the murder of the Rwandan Prime Minister and her ten Belgian bodyguards. His arrest is an impartial statement against impunity and a plea that the world not forget the horrors of genocide in Rwanda. If this plea is not heard – if Mladic and Karadjic are still free by the end of 1996; if the UN denies the proper resources to the Rwandan Tribunal; if the world does not try and build on these two experiments to set up a permanent criminal court – it must not be for want of trying.

The two international tribunals are critically important, but they need to do more than hold a handful of high-profile trials in Arusha and The Hague. These tribunals must persuade all governments to take on prosecutions. Their first trials must lay the foundation for an open-ended, international hunt for war criminals such as was launched at Nuremberg after the Second World War. Second, we must move from *ad hoc* tribunals that are set up to respond to one-off cases of genocide and war crimes to a permanent criminal court capable of applying the same standards fairly and universally.

In the end, it comes down to political will – that other familiar slogan. Historians will conclude that the UN Security Council was paralysed during the Cold War. To judge from Rwanda, Somalia, Bosnia, Liberia and Chechnya, they may draw the same conclusion about the Council in this post-Cold War era. The Council shrank from sending peace-keepers during the genocide in Rwanda; shrank from confronting the Bosnian Serbs during the first three years of war; shrinks from confronting the warring factions in Liberia; shrinks from confronting Russia in Chechnya. It lashed out in Somalia, killing hundreds, and left in a funk.

Until the Great Powers can rediscover their nerve and find a more efficient and legally acceptable method of intervention, the warlords and killers will continue to act with impunity; human rights and humanitarian law will be flouted; and civilians will be denied the protection to which they are entitled.

7

LIBERIA
CAN RELIEF ORGANISATIONS COPE WITH THE WARLORDS?

Fabrice Weissman

•

The first refugees began moving into the Mamba Point sector of Monrovia shortly after 10 in the morning on 6 April 1996. Echoing in the direction of the Sinkor district on the way to the airport came the crackling of automatic rifle fire and the dull thud of rocket-propelled grenades. By late afternoon, civilians by the thousands, as well as some peace-keeping personnel, carrying personal belongings and food tied up in bundles, fled as columns of black smoke rose from different parts of the Liberian capital.

For the residents, Mamba Point had been long considered the most, if not the only, secure part of the city. There, most of the international relief agencies and the few remaining diplomatic missions, among them the United States embassy, the European Union representation and various United Nations offices, were based. Nonetheless, it had remained a highly precarious safe haven, dependent on the will and ability of the international community, the West African peace-keeping forces, and the Americans to assure its protection. As humanitarian relief workers prepared to evacuate their own offices, abandoning vehicles, computers, and radio transmitters, throngs of Liberians crowded into Greystone, a US residential complex a few hundred metres from the embassy, while others sought refuge in and around the aid organisation compounds.

Following days of growing tension as armed factions of Liberia's rival warlords jostled for position in a tense standoff, heavy fighting broke out once again. The conflict-ridden nation, which only months earlier had appeared on the verge of re-establishing peace, was thrown into a renewed state of violence. And, as usual, it was Liberia's war-weary civilian population who suffered.

A SLIM CHANCE FOR PEACE

And yet, at the very beginning of 1996, Liberia seemed to be heading on the path to peace. On 19 August 1995, the main faction leaders had signed in Abuja a 13th peace agreement, which was meant to put an end to six years of civil war. Unlike earlier arrangements, that signed in Nigeria

allowed one to hope for a peaceful settlement of the conflict. All the military factions involved in the fighting agreed to its terms. The National Patriotic Front of Liberia (NPFL) of Charles Taylor – which had initiated the upheaval in 1989 – had signed the compromise along with both branches of the United Liberation Movement for Democracy in Liberia – ULIMO-Johnson and ULIMO-Kromah – the Liberian Peace Council of George Boley and the former Armed Forces of Liberia (AFL) of General Bowen.

Moreover, Nigeria, which had prevented Charles Taylor from taking over Monrovia through the force of interposition set up by the Economic Community of West African States (ECOMOG), no longer seemed to oppose the rise of the NPFL to power. In June 1995, Charles Taylor had visited the Nigerian capital, where he supposedly struck a deal with the junta: the latter would support his political ambitions in exchange for economic advantages over Liberia's natural resources. Backed by the USA, the UN and the 16-nation Economic Community of West African States (ECOWAS), the agreement sought to force the country's rival factions to accept a sweeping

Map 7.1 Sierra Leone and Liberia: Médecins Sans Frontières project sites are situated at Freetown, Kenema, Bo, Kambia, Pujehun, Kailahun (north-east of Pendembu) in Sierra Leone and at Monrovia in Liberia.

Sierra Leone: empowering the women (and society)

For the outsider, the name of the movement comes across as unrealistically optimistic given this country's past three decades of corruption and economic devastation. But then, the acronym does come straight to the point: WOMEN – Women Organised for a Morally Enlightened Nation. And why not? No one else has succeeded, so maybe it is high time that the women started running the show.

WOMEN emerged in 1995 as a key force for pressuring Freetown's military rulers to hold free and fair elections, and then step down in favour of the newly elected civilian government. The woman behind WOMEN is Mrs Zainab Bangura, a former insurance underwriter who has dedicated herself to running what has become one of the country's single most influential organisations. A forceful and exceptionally persuasive individual, Bangura makes it clear that she has no intention of allowing the movement to sit back on its laurels. 'I am very concerned about the months ahead. This war must stop and we must get on with building a new society. And that includes a real democracy which is what I think people here really want,' she argues.

Even though Sierra Leone is now run by a civilian administration, Bangura has very little patience for the discredited and corrupt politicians who have managed to slip their way back in. 'Many of those now in parliament are former politicians with no credibility. Nothing has really changed, so it's now up to the women to take on the government.' Only three out of the 80 new parliamentarians are women. Bangura reckons that women should constitute at least half, including cabinet posts. If the men continue to deny them fair representation, Bangura warns, they intend to turn WOMEN, currently a non-partisan movement, into a women's political party. Only in this manner, she says, can they keep both the parliament and the executive sufficiently focused on the principal issues needed to put Sierra Leone back on the road to recovery.

The story behind WOMEN is indicative of the anger and frustration that has emerged over the years among the country's grassroots. Supported by an initial $25,000 grant from the US Agency for International Development (USAID), Bangura's first expense was to purchase a laptop computer. She was also given two modems. Bangura first became aware of what women could do when she made a six-week visit to the USA as part of the Pluralism and Grassroots Democracy programme organised by the American embassy. There she met with numerous groups and individuals, including women's rights activists as well as civil and human rights advocates.

On her return to Sierra Leone, Bangura and her supporters began working with existing grassroots groups, such as the market women in Freetown. 'We looked at the effects of war on our society, what it was doing, and what we needed to stop it.' Bangura's efforts produced encouraging results. Growing numbers of women embraced her objectives. In 1995, donors such as Great Britain and the USA began putting heavy pressure on the military to hold elections. The soldiers responded by agreeing to appoint James Jonah, a Sierra Leonean and a former senior UN diplomat, as head of a National Electoral Committee, which would organise a consultative electoral conference. The women started attending political meetings. When it was announced that women would receive only four out of the 164 delegate seats at the convention, they protested to Jonah, who managed to have the number raised to 17.

Realising that the women had become a powerful lobby, Jonah used them to keep up his own pressure on the military. But the soldiers, who had formed their own political party, were in no hurry to relinquish power. Arguing that peace should come before elections (an approach which might have kept them in power for years), they sought to buy off delegates. Angered by such blatant vote-rigging, the women announced a pre-conference demonstration. Thousands of women, and men too, marched through the streets demanding elections, despite violent efforts by the soldiers to stop them.

Pressure by the women eventually helped force the military authorities into holding two-round elections in early 1996. It also enabled Sierra Leoneans, both men and women, to become more aware of the democratic process. As far as Bangura is concerned, it is up to Africans themselves to take political matters into their own hands. 'The attitudes of people have changed dramatically,' she said several weeks after the elections. 'They have become more demanding in what they expect from government. But most people, including parliamentarians and ministers, don't know how to go about implementing democracy. It is a massive task.'

peace plan aimed at demilitarising the Liberian capital, disarming the war's estimated 40,000–60,000 fighters (many of whom were in their teens), returning all equipment stolen from the relief agencies, and preparing for nation-wide elections a year later. A Council of State was formed, comprising three warlords and three civilians. The true reins of power were held by the Council's three factional leaders – Charles Taylor (NPFL), George Boley (LPC) and Alhaji Kromah (ULIMO-K). All attempted to protect their respective interests, both political and commercial, in the capital and in the countryside. According to Amos Sawyer, a former interim president, the transitional government created under the Abuja accords was completely taken over by the three warlords, who have consistently sought to crush all forms of civilian opposition. One of the problems, too, was that the council had failed to include the four other leaders signatory to the agreement, most notably Roosevelt Johnson, a breakaway ULIMO rival.

Initially, the Abuja peace initiative appeared to be a serious attempt to encourage, if not enforce, the sharing of power. The factions committed themselves to follow the Abuja principles. In practice, however, the overall situation remained factionalised, with little effective cooperation. If anything, power polarised around Charles Taylor. What the initiative failed to recognise was that this war has been largely fuelled by the trading of Liberia's natural resources for arms and cash. The faction leaders are akin to genuine political–military entrepreneurs with extremely organised clandestine networks backed by local and outside business interests. British, French and other foreign firms have provided a very reliable source of support, notably for Taylor. Liberian wood exports have more than doubled since the launching of hostilities; diamond exports have reached 220,000 carats in 1995, while rubber marketing and production have never ceased. There is no doubt that safeguarding the dividends of this clandestine economy constitutes, for the faction leaders, one of the main stakes of the conflict. Hence the difficulty in installing a peace process which will inevitably reconfigure the commercial interests of the various leaders.

The ethnic dimension of the conflict has also been raised as one of the obstacles to the proper implementation of the Abuja agreements. In fact, this journalist as well as other authors can testify to the fact that many civilians killed during the war were deliberately murdered on the basis of their tribal origin, supposedly for collaboration with the enemy. Yet it needs to be strongly emphasised that the conflict does not result from atavistic tribal hatreds that would naturally thwart any peace process. In an article published by the journal of the High Commissioner for Refugees, an American researcher writes:

> Historically, there has been very little ethnic strife in what is now Liberia. Liberians of completely different ethnic groups lived side by side for decades, and intermarriage between ethnic groups was common. . . The abuses that occurred between the various ethnic groups during the civil war should therefore not be seen as inevitable consequences of long-standing ethnic hatreds. Ethnic difference only became a serious problem with Doe's accession to power, his privileging of the Krahn, and his severe mistreatment of Gios and Manos.

It is indeed the successive monopolisation of the state and the economy by factions artificially built upon ethnic grounds that has given tribal belonging its political and economic importance. Ethnicity has only become an element of Liberian political life as the end result of a history that stretches back to the founding of the country by freed American slaves. Settling in Liberia in the nineteenth century in order to create a Land of Freedom, the Americo-Liberians, considering themselves superior to the native Africans, ran the country in much the same manner as European settlers elsewhere in the continent. During the following 150 years, these settlers imposed a Black-over-Black model of colonization, dominating the country and, as a ruling class, exploiting its resources exclusively for their own gain. The confiscation of power and wealth by Americo-Liberians has generated feelings of resentment on the part of the indigenous population.

Exploiting this hostility, a native, Sergeant Master Samuel Doe, seized power through a coup in 1980. He was, at the time, surrounded by men of Krahn, Mano and Gio origins, but also by Americo-Liberians, such as Charles Taylor, who headed the General Service Agency until 1983. No liberalisation, however, would follow the overthrow of the regime. Instead, with the firm support of the USA until 1985 (Liberia was at the time the African country that benefited from the largest amount of US assistance per capita), Doe progressively destroyed everything around him, and grounded his despotic authority on dependable men. Relying upon the inhabitants of his village, and then upon his own ethnic group (the Krahn), he eliminated little by little the Gio, the Mano and the Americo-Liberian elements whose loyalty he doubted. At the same time, he started a process of *rapprochement* with the Mandingo, while engaging, in Nimba County, in a policy of violent repression against the Gio and the Mano, whose dissident ambitions he feared. In 1985, a coup attempted by an officer from Nimba was followed by bloody reprisals against the county's population, causing between 400 and 2,000 casualties.

In 1989, when Charles Taylor unleashed an insurrection from Ivory Coast, Samuel Doe's forces responded with renewed violence in Nimba County. This encouraged the Gio and the Mano to join the NPFL. A year later, Doe was captured and killed by a group of rebels. By then, however, a vicious civil war had erupted in Liberia. But the ethnic dimension of the massacres which occurred is in no way the product of age-old tribal hatred, which would inevitably doom to failure any attempt at peace-making. Rather, it is the result of contemporary political strategies which attempt to conflate power struggles with issues of identity. As such, these strategies could have led to a political settlement of the conflict. Unfortunately, in April 1996, peace-making did not seem to be part of the warlords' agenda. And, as such, political and economic factors, rather than tribal mentalities, account for the resumption of the hostilities.

BACK TO WAR

Despite unconvincing claims made right up to the renewed April violence by the American embassy, the UN Observer Mission in Liberia (UNOMIL) deployed in the country since 1993, and others, that the peace plan was still working, it was clear that it had failed. According to the accords, the country was supposed to have been disarmed by January 1996. Instead, Taylor (with the apparent connivance of ECOMOG) and also Alhaji Kromah and Roosevelt Johnson, were infiltrating the capital with both men and weapons. Only two days prior to the fighting, John Langlois of the Carter Center, whose Atlanta-based institute has been trying to help the Liberians prepare for elections, declared: 'The peace accords are dead and have been for quite some time. The factional leaders were never interested in giving up their control, whether in the ministries or with business. There are too many commercial interests involved.' This view was shared by other observers, including many of the relief agencies.

The war erupted when Taylor and Kromah joined forces, accused Johnson of derailing the peace process and of murder(!) (a charge that could apply to any of the factional leaders), and demanded that he and his Krahn fighters surrender. Following a weeklong standoff, the shooting started on 6 April near Johnson's house in the Sinkor suburbs. The fighting, followed by looting, soon spread to other parts of the capital. Possibly 200–300 people died in the actual fighting, but many civilians were killed or injured by stray bullets. Groups of armed fighters, often most interested in using their RPGs (Rocket Propelled Grenade launchers) for looting, roamed the city. The ECOMOG peace-keeping forces, disappearing from their checkpoints and other positions, made little or no effort to halt the spread of fighting or looting. Johnson and thousands of his supporters, including women and children, were holed up under appalling conditions with little food, water and almost no sanitation in the Barclay Center, a military training facility located between Mamba Point and Sinkor. The USA sent in its troops and helicopters to evacuate over 2,000 American citizens, international relief workers, missionaries, Lebanese businessmen and their families, and to rescue expatriates stranded in different parts of the capital. The US government, however, made no effort to interfere in street fighting or to protect vulnerable Liberian civilians.

The resumption of hostilities was an occasion to witness the behaviour of the various factional fighters. Most of them are marginalised teenagers who have broken with the traditional social order and have failed to integrate themselves into urban society. Instead, they have found in the life of a warrior both a trade and a source of status. These fighters, largely addicted to drugs, display little sense of discipline. They terrorise local populations, using their weapons to assert themselves, whether at checkpoints or in inhabited areas where they steal food, force local inhabitants to work for them, and rape women. In addition, they make work particularly difficult for relief agencies. The latter are constantly subject to the pressure of these young men, who do not always abide by the orders of their officers and seek to divert part of the humanitarian aid for their own benefit.

The resulting situation, however, is far from being totally chaotic. There exists a minimum of order in these factions made up of autonomous battalions which are, in turn, divided into a multitude of combat units. The control over the flow of arms and ammunition, the marketing circuits of a prosperous clandestine economy, the use of magic, and the eternal game of

Plate 7.1
Child soldier with his weapon in Liberia. Many children with guns know no other way of life.
Photograph: Jan Banning

Extract: A letter from Sierra Leone

Dear Tout Le Monde en Nouvelle Zealande, Australie, Angleterre, et les Etats Unis

How to explain the situation here? The basic military mosaic I think you have, but to refresh; there is the Government military, the RUF, SOLBELS (soldiers who become rebels), ULIMO (mercenaries from Liberia), the South Africans (for the DeBeers Mafia/Cartel), the traditional hunters called Kamajos, the civilians; there are also bandits etc. etc. Add to this a recent coup d'état and elections, one can begin to comprehend the confusion that abounds here. There are also the remnants of cannibalism, voodoo, etc. to muddy the picture.

I was stuck in Bo during the main elections. It was very bloody. Each day we received about 20–30 war victims in various states of mutilation. Some with amputations of the hands to prevent voting and RUF, TERROR, NO ELECTIONS tatooed across bodies, plus amputated lips, ears, for what reason? Je ne sais pas.

The interesting thing is that there has only been a marginal decrease in the amount of amputations etc. after the elections. Recently I received a man with . . . his ear (cut off) which he was made to eat!!!

There is a myth that I have seen in the press that there is only mutilation and very little killing. . . . For every victim I receive at the hospital at least 3–4 are dead. Not just assassinated but eviscerated, beaten, babies cut up, torturing etc. We just see the tip of the iceberg. We see so much the iceberg must be very big indeed.

During the elections it was psychologically very arduous. There was massive retaliation by the civilians against the military. There were crowds of youths with machetes and sticks. Every time the crowd roared and surged we knew someone would be decapitated and dismembered. When I could get back to the hospital after the mortar fire had diminished I went to the morgue to do an inventory of the dead. The first bodies I saw were wrapped in shrouds – no problem. Then the attendant opened another door and I saw four corpses without heads but with one head to the side. He said I have more like that. . . . Anyway the whole thing here is very macabre.

Now there is supposed to be a cease-fire between the rebels and everyone else. This is utter bollocks. The military continue to do the most brutal things to the general populace. Recently we had a man come in post election with a bilateral hand amputation. No sooner had he arrived than the local military turned up to take photos for the press saying that the rebels had broken their cease-fire agreement. We all know that it was the local brigade that had instigated this. One day after this attack MSF and AICF went to the area and the local military had said there was no problem pertaining to rebel activity!

Sandra Simons, MSF nurse from New Zealand

divide and rule allow the faction leaders and the commanders to enjoy a certain degree of authority over their men. This can be seen in the fact that their vital interests are indeed protected, that their war economy does function, and that a certain number of buildings in Monrovia have been systematically spared from the looting. Even at the height of the fighting, the brewery never stopped producing its daily output of beer. The offices of the European Union representatives, as well as the headquarters of the former French embassy, have never been touched. Neither has the centre of telecommunications nor the Mamba Point Hotel, where most of the international journalists stayed. Faction heads certainly did not want to antagonise donor countries, the world press, or be deprived of telephone lines.

By May 1996, West African leaders with American and European support tried to persuade the Liberian warlords to agree to a cease-fire. Johnson himself was hustled out of the Barclay Center in a US armoured convoy to attend the peace talks in Accra, Ghana. Taylor refused to participate. Instead, demonstrating his lack of interest in a political settlement, he dispatched a representative. This may have proved a serious miscalculation. Relying on his belief that Nigeria, but also many Liberians, would be prepared to accept him as a head of state (if only, for the latter, to put an end to the

Plate 7.2
A young victim of a rebel attack is brought in for treatment in Bo, one of the main destinations for internally displaced refugees. Photograph: Kadir van Lohuizen

war), Taylor may have thought that the April events would speed up his accession to power. By the end of May, despite little prospect for an immediate end to the conflict, the USA seemed determined not only to ride out the crisis but to step up its pressure against Taylor. ECOMOG, too, was beginning to reposition itself in the capital and impede the movements of his NPFL fighters. And, in addition, Johnson's fighters were beginning to score against Taylor's. Incapable of overcoming Krahn resistance, the NPLF and ULIMO-K government forces agreed, on 26 May, to cease fighting and let ECOMOG be deployed in the capital. In early July, while Monrovia was quiet again, hostilities continued and intensified in the rest of the hinterland. This forced Liberia's civilian population to seek refuge in the neighbouring countries or in the highly fragile security zone that had been restored in the capital.

At the time of writing, it is clear that the entire peace process will need to be fully reassessed. This includes a complete reappraisal of the roles and intentions of the different players, ranging from the factional leaders and their commercial supporters to ECOMOG, the Americans and the European Union. For the international aid community, in particular, guarantees of security for civilians and also for aid workers, their equipment and supplies will determine the nature, extent and justification of humanitarian relief operations.

LIBERIA'S CIVILIANS: PAWNS IN THE GAME

It was the US evacuation that drew international attention (for a short time at least and only because Americans and Europeans were involved) to the Liberian conflict. Yet it was the exodus from Monrovia in early May 1996 of over 2,000 Liberian refugees and other African nationals, including 26 ECOMOG soldiers, aboard the leaky Nigerian vessel *Bulk Challenge* followed, like a shadow, by the Médecins Sans Frontières ship *Orca*, that focused attention on the plight of these increasingly desperate civilians. Trapped with little food and water, the refugees were, in addition, subjected to highly unsanitary conditions.

This exodus underscores the distress of the local population. Throughout the conflict, civilians have been the victims of much abuse. Liberians who have not succeeded in escaping into the bush, thereby evading the authority of the warlords, are enslaved by the military groups. Often, they are used as human shields during military attacks, are forced to labour for the various factions or to give up their children who are then forcibly conscripted into the ranks of the fighters. At times deliberately starved, the civilians are used as lures for medical assistance and food supplies subsequently co-opted by the battalions. This is why displacements of populations are so considerable. According to the World Food Programme, today more than 80 per cent of Liberians are displaced or refugees. According to UNHCR, more

Plate 7.3
Government soldiers
on patrol in Sierra
Leone. Photograph:
Kadir van Lohuizen

than 750,000 Liberian refugees have sought a safe haven in other countries
in the region, mainly Ivory Coast and Guinea, but also Sierra Leone, Ghana
and Nigeria.

But the story of the *Bulk Challenge* also illustrates the increasing reluctance of bordering countries to welcome new Liberian refugees. The Ivory
Coast, which already sheltered some 300,000 Liberian refugees, maintained
that it could not possibly take any more. Contending that at least 2,000
on board were combatants, an allegation flatly denied by the relief agencies (over 1,000 were women and children), the Ivorian authorities finally
allowed the ship to dock for emergency repairs at San Pedro. Women and
children were permitted to disembark and were transferred to nearby warehouses, where relief workers provided them with food and medical
assistance. Two days later, they were ordered to return to the boat. The
Bulk Challenge was eventually allowed to discharge its human cargo at
Takoradi in Ghana. At the same time, nearly 1,000 other refugees were
undergoing a similar ordeal on board the fishing vessel *Victory Reef*. After
considerable international pressure, they were permitted ashore in Sierra
Leone after six days drifting off the coast. The 400 passengers on board the
Zolotitsa were not so lucky. Unlike the *Bulk Challenge*, whose voyage was
monitored and publicised by MSF, the plight of this dilapidated Russian

fishing boat, which floated at sea for more than three weeks, was treated with indifference. After having been forced back from both Ghana and Togo, the passengers, who had paid up to $100 for a passage to Accra, finally landed in the port of Monrovia.

HUMANITARIAN RELIEF: HOLDING ON OR PULLING OUT?

As one of the world's 30-plus conflicts in 1996, Liberia represents an increasingly current predicament whereby millions of civilians are caught up in a ruthless and enduring war. Over 150,000 Liberians have died in nearly seven years of war. At least half a million have been displaced as internal refugees within their own country. Overall, well over two-thirds of Liberia's pre-war population of 2.6 million have been forced to flee their homes. The country's health system has collapsed, while local production of rice, which even before the war had been incapable of providing for the country's needs, had gone down, in 1995, to a mere 23 per cent of its 1989 level. In situations such as this, humanitarian agencies are often the last barrier between basic survival and death for civilian populations.

At the same time, scores of veteran aid agencies are finding themselves in the awkward position of inadvertently aiding the survival of militia and armed factions. In six years of war, Liberian warlords have become masters in the art of diverting relief assistance: creation of starvation to attract medical and food relief that is then partially diverted; channelling of assistance to provision armed groups and to control civilian populations; imposition of taxes and racketeering at road checkpoints; the need for armed guards, paid in supplies, for convoys; attacks on the very same convoys; thefts from the warehouses; taxation of the population after the distribution of food supplies; forged lists of beneficiaries ... This is without mentioning the fact that since the end of 1991 every offensive has been accompanied by an almost total looting of the aid agencies in the areas concerned. In September 1994, during the course of the offensive of the anti-Taylor factions against the NPLF capital, Gbarnga, aid agencies lost more than 5 million dollars' worth of food supplies and equipment, including around a hundred vehicles.

This looting was not followed by any concerted response by the relief community. Except for a few agencies – such as the International Committee of the Red Cross (ICRC) – most assistance organisations resumed their activities as though nothing had happened. Donors did not make use of their leverage *vis-à-vis* the faction leaders to impose some minimal respect for humanitarian principles. A kind of *carte blanche* was delivered to the warlords, who became respectable negotiators at the international round table. In August 1995, MSF, in cooperation with DHA (Department for Humanitarian Affairs), elaborated a document defining a certain number of principles of intervention that aimed at imposing upon the factions respect for a humanitarian domain. But this initiative was not realised. The multiplicity of relief agencies and the lack of coordination between them made it impossible to impose upon armed movements a common front likely to limit the many violations of humanitarian law.

The same scenario unfolded once again during the April events. International relief teams in Monrovia (as well as those operating in other parts of the country) lost an estimated $20 million worth of equipment, including at least 200 vehicles. The breakdown of law and order prompted virtually all the aid agencies to shut down their operations and evacuate their expatriate personnel. Efforts to remove or paint over logos before abandoning their compounds were ineffective. Four-wheel drives and trucks sporting the logos of MSF, UNICEF (the UN International Children's Emergency Fund), UNHCR (UN High Commission for Refugees), Save the Children and others therefore became more closely identified with the armed factions than with medical relief and food distribution. After the resumption of hostilities in April 1996, some of the agencies considered maintaining basic minimum staff to ensure a humanitarian presence in the field, if only to show the civilian populations involved that they were not totally abandoned. MSF and the ICRC, for example, jointly established an emergency medical centre to handle injured civilians. But eventually they were forced to withdraw. With considerable reluctance, many of the agencies left behind several thousand local Liberian staff. Only a small number of UN personnel remained deployed at a secure base north of Monrovia. But there was little they could do to assist the tens of thousands of Liberians, who, once again, had been forced to flee, many of them losing all that they owned.

Perhaps more than any other recent crisis, Liberia has raised the question of whether international relief agencies should continue with humanitarian operations without the back-up of appropriate security. Should relief workers seek to save lives in the short term but risk exacerbating the conflict by provisioning the warring factions with food, vehicles and supplies as well as other sources of equipment and revenue, thereby condemning more people to deprivation and death in the long term? Or should they stand by and do nothing because more lives may (but also may not) be saved eventually? Is this a decision for humanitarian agencies? Or has the onus fallen, once again by default, on the aid organisations to find the solutions (when obviously they cannot) because the international community has failed to provide the necessary political or military support for them to operate?

Increasingly debated by most of the serious relief agencies, such options place ICRC, Save the Children, MSF, Oxfam, UNICEF, UNHCR and others in a bind. 'International aid representatives are not necessarily suggesting that the humanitarian aid organisations are making matters worse. But it is a question we constantly ask ourselves,' observed one UN official in

Plate 7.4
A rebel soldier in
Sierra Leone.
Photograph: Kadir van
Lohuizen

Monrovia. According to the ICRC, the Swiss-based organisation was frequently accused of prolonging violence in Somalia because, based on its own estimates, up to 5 per cent of its food aid was being misappropriated by the warlords. Nevertheless, it argues, 1.5 million Somalis survived the famine thanks to international humanitarian assistance.

Faced with the prospect of trying to help beleaguered civilian populations at the risk of further aggravating the crisis, most relief workers have convinced themselves, or are at least trying to convince themselves, that continued humanitarian relief to Liberia remains imperative. Few are willing to declare openly that letting Liberians fight it out amongst themselves may be the only way to deal with the conflict. While Liberians themselves admit that they bear the responsibility for their own predicament and should not expect the outside world to bail them out, most relief workers feel that the international community should assume a greater degree of responsibility in helping to solve the crisis (at least in fighting against the illicit trading which serves as a financial basis for factional war efforts, or in imposing sanctions against those faction leaders responsible for the resumption of hostilities) or at least helping the civilian populations to find refuge in genuine security zones inside the country or in neighbouring countries.

Relief agencies view the restoration of effective security – at least in a

few specific areas where they could efficiently monitor the distribution of their assistance, and where they would not face every day the risk of seeing their equipment looted – as a basic condition and the only feasible way of overcoming this dilemma. Since the failure of the UN operation in Somalia, however, the USA and other countries have been reluctant to support the sort of military operation required to create genuine security zones in politically volatile states such as Liberia. Humanitarian aid has been used as a substitute for the political commitment required to ensure minimum protection for the civilian populations trapped in the conflict. 'What has happened in the last few years is that humanitarian aid has become a tool that governments use to avoid real and concerted action,' asserted MSF's Steve O'Malley in New York. Similarly, while the principal donors are fully aware that regional peace-keeping without proper accountability has not worked, and cannot work, as in the case of ECOMOG, the international community seems to persist with the charade.

Plate 7.5
Each day, medical relief workers such as this MSF volunteer in Kenema, Sierra Leone, treat victims of all ages. Many have been wounded by soldiers or rebels, often in revenge or as a warning to others. Photograph: Remco Bohle

ECOMOG: PEACE-KEEPERS OR PROFITEERS?

For many observers and humanitarian representatives, the West African peace-keeping forces, particularly the Nigerians, who represent 66 per cent of the soldiers, there is little question that ECOMOG has contributed to the conflict. This regional force is very much a part of the Nigerian military regime's own foreign policy and commercial interests. Encouraged by the USA and the UN – within the framework of their policy of delegating peace-keeping operations to regional organisations – the deployment of ECOMOG is not simply the result of the collective security mechanisms provided for by the ECOWAS. The force of interposition is first and foremost the product of Nigeria's strategy to stop the advance of the NPFL. The latter was supported by Libya and Ivory Coast – both countries which challenge Nigeria's claims in the sub-region – and seemed poised to destabilise the military regimes allied with Lagos. Concealing their anti-Taylor agenda behind the cover of the ECOWAS, the Nigerian battalions of ECOMOG have from the start appeared as a partisan force. Not only were they allies of the former government forces to thwart Taylor's offensive on Monrovia in October 1992, but they also took part in the creation of the LPC, which they regularly supplied with arms and ammunition. In 1993, ECOMOG imposed a humanitarian blockade on Taylorland, not hesitating to shell civilian hospitals, food storage facilities and aid convoys. Moreover, the Nigerian White Helmets have taken an active part in a clandestine economy. Since its intervention in 1990, ECOMOG has been responsible for a large part of the looting, whether it was directly involved in pillaging or in its organisation, or else in buying goods from the looters. Officially known as the ECOWAS monitoring group, ECOMOG has become more appropriately known as 'Every COmmodity MOvable Gone'. Over the years, its troops have made a concerted and unabashed effort to remove anything of value, ranging from cars to roofs, piping, to diamonds, shipping most of it to Nigeria. Early in 1996, fighting even erupted between ULIMO-Johnson and ECOMOG in Tubmanburg when a new Nigerian commander refused to pursue collaboration on a diamond-mining arrangement worked out by one of his predecessors. The Nigerians have been similarly involved with faction leader George Boley in exploiting rubber resources. But ECOMOG soldiers have also gone into business for themselves by running their own operations.

According to sources, at least six former Nigerian commanders have developed substantial business operations in Liberia and are still using their ECOMOG contacts to continue dealing. The *Bulk Challenge*, for example, was reportedly carrying vehicles and other commodities looted during the April fighting, many of them sold directly to the Nigerians at the ECOMOG base.

One must, however, give credit to the White Helmets for maintaining, between 1990 and 1995, a relative security zone in the government areas. The capital until 1992, and the Monrovia–Kakata–Buchanan triangle starting in 1993, have constituted a safe haven, where civilian populations fleeing the clashes and abuses of the fighters have been able to take refuge.

Those areas were the only ones which were regularly supplied with international assistance and where aid agencies were able to work in relative tranquillity. For this reason, at the beginning, many Liberians were ready to view the plundering of their country as a price worth paying for security. But anger started growing as people realised that ECOMOG not only took advantage of Liberia's plight for its own benefit, but also failed to provide adequate security.

Indeed, during the April violence, the peace-keepers made little effort to interfere and protect civilian populations. Apart from a number of operations to rescue people such as aid expatriates, UN local staff and their families, and (at a price) Lebanese businessmen, they abandoned their checkpoints almost as soon as the fighting began. Had they remained at their posts in a show of force, they might have been able to contain at least some of the fighting. ECOMOGs supposed impartiality became even more questionable when it appeared that it had overturned its earlier alliances and made a deal with Taylor, allowing him to infiltrate fighters and weapons into Monrovia during the weeks that led to the fighting. While the West African community claims, with some justification, that the force is too poorly equipped and badly trained to perform an effective role, it is more than that. This force has proved to be too partial and too involved in clandestine economic activities to carry out its duty.

At the time of writing, new peace initiatives call for a strengthening of ECOMOG. The USA has offered 30 million dollars in support, which would include new equipment and training. Belgium and Denmark have offered to pay for additional peace-keeping battalions from Ghana and Burkina Faso. According to the Ghanaian deputy foreign minister Mohammed Ibn Chambas, a 12,000-man force, which would cost $90 million a year, should be provided by the international community. Instead, he complained, foreign countries prefer to spend large sums of money on humanitarian assistance while the roots of the crisis remain unaddressed. While European and American donor representatives agree that such a force needs to be properly equipped in order to be effective, some question whether ECOMOG, given the agenda of Nigeria's military dictatorship and its obvious desire for expansion in the region, is the right vehicle to help implement the peace process or even to guarantee the integrity of genuine security zones. As Colin Scott, a British relief consultant, observed in a 1994 report for the Thomas J. Watson Centre for International Studies, the West African peace-keeping operation had clearly stepped into a political vacuum but had failed to maintain a neutral stance. It also raised the question as to whether a UN intervention force might not have produced a better long-term outcome. For this reason, observers suggest that Liberia would be served better by involving troops with no particular affiliation or interests in the region, such as Eritreans or Pakistanis, or by placing the ECOMOG forces under UN, or even American, control. None of the faction leaders, or most Liberians for that matter, will ever really trust ECOMOG to police a reinstituted peace settlement or to protect security zones as long as its commanders and soldiers retain a personal stake in the country's commercial and political future.

THE CHILDREN OF AMERICA: WHOSE CHILDREN? WHOSE RESPONSIBILITY?

Soon after the April events, West African leaders expressed disappointment that the USA was unwilling to become more directly involved. Other than providing relief shipments, some financial aid and support for the evacuation of American and other foreign nationals, Washington has held back. This was made clear during the Gulf War, when Liberians failed to understand why American troops were prepared to intervene for the sake of oil-rich Kuwait but not to help save lives in backwater Monrovia. At the time, observers felt that a relatively small US show of force in the Liberian capital could have helped prevent some of the killing.

Given the USA's long-term affiliation with Liberia, which was founded in 1847 as the continent's first black republic, many Liberians like to refer to themselves as Children of America. It was the USA along with the American Colonization Society, that helped bring over the first boatload of ex-slaves to the shores of Liberia in 1822. Many Liberians feel, that if any African country has claim to being the focus of US attention, it should be Liberia. Several hundred thousand Liberians or Americans of Liberian origin live in the USA. There is little question that numerous Liberians would like nothing better than for the Americans to intervene.

Political realities in Washington, however, dictate that such intervention is highly unlikely, especially since the failure of the operation Restore Hope in Somalia. Liberia is simply too remote to be considered part of America's vital interests. Since the sale of the Firestone gum plantation to the Japanese firm Bridgestone, the USA no longer has any major economic concerns in the country. Their Omega submarine station has been transferred to Mozambique, and the retransmitter of Voice of America has moved out. And the Liberian conflict does not figure prominently in the Black Caucus political agenda. 'We have had a great many black Americans tell us over the past few years how 30 million African-Americans are behind us, but these are just words. They really don't care about us,' said a Liberian nurse who trained in the USA. Finally, many observers feel that any form of American intervention in Liberia would be disastrous.

TO RECREATE A SPACE FOR HUMANITARIAN AID

If it is difficult to imagine an international intervention which could solve the Liberian conflict, it is nevertheless possible to think about the manner

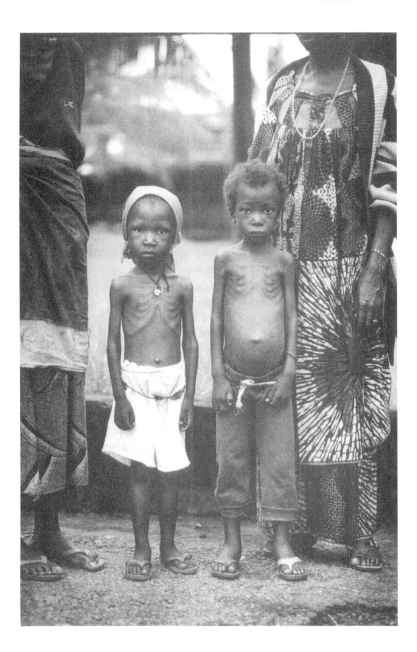

Plate 7.6 Malnourished children at a relief agency feeding centre in Kenema, Sierra Leone. Food aid targeted at civilians often ends up in the hands of government soldiers or rebels. Photograph: Remco Bohle

in which aid operations could acceptably be resumed. One thing is certain: relief agencies cannot resume their activities in Liberia under conditions similar to those which have prevailed during the last six years. The role inadvertently played by humanitarian aid in the factionalised war economy and in population control calls for the elaboration of new methods of intervention. The legitimacy of all future operations is at stake.

First of all, it seems imperative to limit the risk of the diversion of aid for military purposes. Relief agencies must therefore dramatically reduce the extent of their own logistical investments. Considering that factions show little respect for relief agencies, the latter must refrain from introducing trucks, cars, motorcycles, radio-communication equipment, computers, destined to fall into the hands of the fighters every time the political–military situation worsens. It is especially important that control be exerted over the distribution of food supplies. They must not be used to feed the factions, or used as a means of enslaving populations taken hostage by armed movements. The implementation of these guidelines will not necessarily enable aid agencies to meet the requirements for humanitarian assistance in every location in which it is needed, particularly in rural areas. This is why all Liberians, deprived of access to relief, should be offered the possibility of seeking refuge along the coast or outside the country, in specially created zones where armed factions would be neutralised. Truly protected by an impartial international force, these secure enclaves would ensure protection for the civilian populations affected by the clashes and violence perpetrated by the fighters, and would allow the relief agencies to work without fearing that their assistance will be turned against its supposed recipients.

Unfortunately, the application of such measures runs counter to the prevailing trend. The creation of genuine security zones inside Liberia, as well as in the neighbouring countries, would imply the deployment of a truly impartial international force, whose decisions would no longer be determined solely by the economic and political interests of the states in the area. Until today, the international community pretended that ECOMOG was capable of undertaking this task. Even when the daily practice of the White Helmets showed that their political and commercial agenda is incompatible with a real mission of protection, the states in a position to intervene continued to close their eyes lest they be forced to become involved militarily or politically. The UN does not appear to be less defensive. In June 1996, the proposal of the Secretary-General to involve the UN more deeply in the Liberian conflict was met by a meaningful silence on the part of all members of the Security Council. The reinforcement of ECOMOG is still on the agenda, while the revision of the peace-keeping operation and the addition of new troops from outside the sub-region have never been discussed.

As the adventure of the *Zolotitsa* testifies, the welcoming of Liberian refugees abroad is not on the agenda either. The Great Powers aim instead to limit the international implications of the conflict by containing the flight across borders of a new wave of Liberian refugees, 750,000 of whom have already sought asylum in Ivory Coast, in Guinea or in Sierra Leone. This is the reason why donor countries encourage an outpouring of humanitarian aid inside Liberia, even if it means deliberately ignoring the hijacking of supplies and the subjugation of populations trapped inside the territory of the various factions.

There is nothing surprising in this practice, which reflects the new containment policies dominating attitudes towards conflicts that are peripheral to powerful countries. Military involvement of regional states, the creation of pseudo-security zones, fed by international assistance, inside the areas of conflict, the closing of the borders, even the forced repatriation of refugees (scheduled for the Liberians in 1996) constitute the pillars of stabilisation initiatives advocated by the Great Powers in recent crises. In Liberia, the inanity of such an approach is obvious. It is pointless, and morally unacceptable, to continue blindly to pour humanitarian aid into the country as long as the civilian population cannot find refuge in real security zones, inside or outside Liberia – safe havens where aid will not be in danger of being used against its beneficiaries. Until it faces up to its responsibilities, the international community could at least impose an embargo upon the private companies which continue to exploit the country's resources for the sole benefit of the warlords.

8

BOSNIA

IN SEARCH OF A LASTING PEACE

Stephan Oberreit and Pierre Salignon

with contributions from Renaud Tockert

•

For Bosnia, 1996 started as the year of peace. The peace, however, raised contradictory hopes for the future. Some thought it would still guarantee a possible united Bosnia-Hercegovina; some that it would mean integration into a Greater Croatia; and others saw it leading to an independent Bosnian Serb state. Such irreconcilable interpretations could not augur well for the future. But the current context of an uneasy balance between peace and possible renewed war needs to be put into perspective. For the last few years, in the heart of Europe, the politics of ethnic cleansing has led to the burning of houses, violent intimidation, shelling of populated areas, sniper attacks on civilians, mass deportations, rape, concentration camps, mass graves and torture, all in order to purify territories from other communities. As in any war, one witnessed also the flourishing of local mafias, criminal economies and militias. Out of a pre-war population of nearly 5 million Bosnians, tens of thousands have lost their lives and 3.5 million have been forced to flee their homes. The issue of protection for this population was widely discussed. Attempts at developing remedial strategies were made under the pressure of international public opinion. On the ground, however, these efforts failed.

It was only in the second half of 1995 that events resulting in the application of force led to a new situation. Until then, strategies which had survived through no merit of their own were revealed as unworkable. The United Nations Protection Force (UNPROFOR) was unable to fulfil even its limited mandate. New approaches had to be developed. By the end of 1995, after the NATO (North Atlantic Treaty Organisation) operations, the balance of power had swung decisively away from the Bosnian Serbs. That, in turn, paved the way for the signing of a General Framework Agreement for Peace in Paris on 14 December 1995. Fighting had at last stopped. Why did this victimised population have to go through such an ordeal for so long before NATO intervened? Why did the allies have to wait for the fall of Srebrenica and more deadly shelling incidents in Sarajevo in order to act? How can one ignore, or conveniently forget, the horrible suffering of a large number of civilians during the past four years in the former Yugoslavia?

ETHNIC CLEANSING IN EUROPE

Remember Vukovar. The Pearl of the River Danube was bombarded, cleansed and, after an 86-day siege, fell into Serb hands on 18 November 1991. An estimated 300,000 Croatian inhabitants fled Slavonia and the Krajina. Thousands more were maimed, executed or killed by the bombs of soldiers serving Slobodan Milosevic, Belgrade's strong man. All this portrayed 'live' before the eyes of the world. Prior to the fall of the besieged city, a convoy of Médecins Sans Frontières had succeeded in evacuating some 100 severely wounded. But it was attacked during this first mission and unable to return to pick up the remaining patients. The international medical organisation was devastated to learn weeks later that all those left behind in the hospital – about 200 wounded – had been executed following the fall of the town, and dumped into pits. Evidently, the Serb troops did not want witnesses.

In the summer of 1992, in the area of Kozarac, but also in the regions of Foca, Visegrad, Srebrenica, Zvornik, Gorazde, Jajce and Sanski-Most, hundreds of Bosnian towns and villages were systematically set on fire and destroyed. Thousands of civilians were taken prisoner and many were

Map 8.1 The Balkan States: Médecins Sans Frontières project sites are situated at Zagreb, Kuplensko (south-west of Zagreb), Split and Osijek in Croatia; Bihac, Jajce, Zenica, Sarajevo, Mostar, Banja Luka, Tuzla, Maoca (north of Tuzla), Maglay, Pale and Gorazde in Bosnia; and at Vojvodina (Novi Sad), Belgrade and Pristina Kosovo in Serbia.

Remember Srebrenica

The memories of the fall of the erstwhile safe haven will haunt Christine Schmitz for ever. Serving on MSF medical staff there with an Australian doctor, Daniel O'Brien, she was forced to witness the brutal expulsion of the civilian population at the hands of the Serb troops.

The MSF team organised the transportation of all the hospital patients to the UN compound at Potocari, provided first aid for the completely exhausted people and helped to deliver several babies during those desperate hours. Last but not least, they succeeded in evacuating Bosnian MSF workers to Croatia – certainly saving the lives of the men among them. Despite the traumas, Christine Schmitz is adamant: 'At no point did I ever wish I were somewhere else.'

How could Srebrenica have fallen into the hands of the Bosnian Serbs when the United Nations had declared the enclave to be a safe haven? What use are all the words and promises, what use are the Geneva Conventions and international agreements on human rights if nobody accepts responsibility for ensuring that they are observed?

When I think back to those days in Srebrenica, these and other questions come to mind over and over again. Powerless and helpless, we were forced to witness the fall of the enclave and the expulsion of the Muslim population from their homes.

Some moments will remain etched into my memory for ever. I'll never forget the joy on the people's faces when, after long days of waiting, a food convoy finally reached the besieged town; the long days in a cold, dark, damp cellar, seeking refuge from the Serb shell attacks; the fear I felt when a shell struck close to the hospital; the screams and the desperation of the wounded who had been brought into the hospital after a particularly heavy attack on the town; how the Muslim population fled the town, in burning heat and under a hail of shelling, in order to reach the UN compound 5 kilometres outside Srebrenica; the young woman giving birth to her child on a filthy stretcher with not even a hint of privacy; how the expellees waited for two days in front of the UN compound with no food, water or shelter while the MSF convoys were being refused permission to pass.

But the flight out of Srebrenica was not the end of the horrors for the expellees. At the time, hardly anyone guessed how well the Serbs had organised the taking of the UN safe haven. In only a few hours, heavily armed Serbs shoved terrified women, children and elderly into buses to transport them into Bosnian territory – towards an uncertain future. One young woman lost control out of pure panic and started kicking and screaming; a weeping father pressed his little daughter into my arms before being led away by the Serbs. This was one of the most terrifying moments for me, since I still didn't know what happened to all those men that the Serb soldiers plucked out of the crowds of refugees.

Besides these terrible memories, I have some experiences upon which I can look back with a certain degree of silent gratitude: the indefatigable readiness of our Bosnian colleagues to care for the injured and the sick despite the intense psychological strain. Or the moment when I was given permission to return to the seemingly empty Srebrenica to fetch the elderly patients I had been forced to leave behind in the hospital days before. The whole time they had waited where we had left them. And the relief when, at 4.00 a.m. on 21 July, after a 16-hour drive, we finally arrived in Zagreb, absolutely exhausted, together with our Bosnian colleagues. Right until the very end, we had not known whether we were going to make it.

The Muslim population in Srebrenica and Zepa seem to have lost their homes for ever. We can only hope that the people in the enclaves of Sarajevo and Gorazde don't meet a similar fate.

Christine Schmitz, German nurse, MSF, Srebenica

tortured. Members of the elite were killed in concentration camps reminiscent of a dark past: Omarska, Manjaca, Trnpolje, Keraterm. The perpetrators were the militias of Radovan Karadzic, implementing a plan that the Serb leaders apparently had drafted with meticulous care. Western countries protested under pressure from international public opinion, shocked by the images of the starving prisoners. The most visible camps were closed, but massive human rights violations and abuses of elementary humanitarian standards persisted. Harassment of the remaining non-Serbs continued. Efforts to force out terrorised civilians succeeded through discrimination and violence. Working hand in hand with the killers, racketeers saw to it that no one escaped with any of their possessions or money. The so-called uncontrolled Serb militias, such as the one led by Arkan, a notorious militia commander, were able to operate with absolute impunity. Western diplomats knowingly sacrificed Bosnia and Hercegovina, silent witnesses to crimes committed in the name of an ethnically cleansed Greater Serbia.

All in all about one million Muslims and Croats were driven out of Serb-controlled Bosnian lands, despite repeated protests from international organisations. By accepting, in some instances, that populations should be evacuated in order to be protected from the violence of the extremists, international organisations have, against their will, compromised with the politics of ethnic purification.

Did they have any other choice?

Despite continuing aggression against unarmed civilians, the international community failed to confront the aggressors openly. Both the United

Nations and the Security Council lacked the political will to impose peace or to give teeth to their various resolutions. Neither the deployment of 40,000 Blue Helmets nor the introduction of 'safe havens' (Sarajevo, Tuzla, Bihac, Srebrenica, Gorazde, Zepa) in 1993 prevented General Mladic's soldiers from attacking Gorazde in April 1994 and Bihac in November 1994. Even the agreed-upon reopening of Sarajevo airport was never implemented. Following each NATO ultimatum, the situation in the Bosnian capital remained calm for a short period, but then the shelling started again.

For the Serb troops, the complacent attitude of the European powers had become a licence to kill. And they were not the only ones to commit crimes. The vicious cycle of violence led all parties to violate the rules of war. The Croats used similar methods to force Muslim and Serb minorities to flee their land: in 1992 and 1993, when they had detained thousands of Muslims in camps around Mostar; and in August 1995, when they caused the exodus of 200,000 Serbs from the Krajina, a Serb-dominated region of Croatia that had claimed independence from Zagreb in 1991. Nevertheless, the misdeeds of the Croats and Muslims, however despicable and inexcusable, never attained the extent and degree of cruelty of the massacres committed by Serb troops in the areas they had cleansed.

THE NATURE OF THE WAR

Bosnia-Hercegovina's push for independence was driven by a well-founded fear of political domination by Serbia. The Bosnian Serbs had similar misgivings about the Muslim-oriented Party of Democratic Action (SDA). Just as the Muslim people of Bosnia had no faith in the governance of Yugoslavia, so the Serb people of Bosnia had no faith in a Muslim-dominated government in Bosnia. The Bosnian Muslims, in alliance with the Bosnian Croats, gained independence in a referendum in February 1992 (boycotted by the Bosnian Serbs) and subsequent recognition by the international community. Thereafter, the leaders of the Bosnian Serbs, in alliance with Serbia, chose force to maintain their domination. Because the Bosnian Muslims, Serbs and Croats were geographically so intermixed when the conflict in Bosnia erupted, it inevitably engulfed the civilian population. In fact it was the civilian population that constituted the very stakes of the war. It was the imposition by force of ethnic will in pursuit of domination that produced the horrors of the war. The simplistic explanation of ancestral ethnic hatreds for the degree of violence encountered is far from convincing. In fact, we witnessed the utilisation of ethnicity by the leaders during the course of a power struggle, demonstrating a conscious use of an available common denominator to manipulate a despairing population in the midst of a multifaceted social crisis.

Plate 8.1
Refugees at the Tuzla
airstrip, Bosnia.
Photograph: Christian
Jungeblodt (Signum)

SPRING 1995 IN BOSNIA

There was little to suggest that 1995 would bring any change in the Bosnian
civil war. Sarajevo remained surrounded and under siege. A cease-fire, negotiated by UNPROFOR in December 1994, brought some respite from
shelling to the city's population, but civilians on both sides of the divided
city continued to die in sniper shootings. It held until March 1995 when
Bosnian Muslim and Croat forces launched attacks on several fronts, in the
Tuzla, Travnik and Krajina regions. These assaults made ground but a consequence was that the peace in Sarajevo ended and the shelling resumed. By
May 1995 the fighting had intensified. The Croats had attacked and taken
Western Slavonia from separatist Serbs. The shelling of Sarajevo had
continued, causing numerous casualties. A determined push by Bosnian
government troops in Visoko, south-east of Sarajevo, was matched by a
typically vigorous Serb response. For UNPROFOR's commander, British
General Rupert Smith, it was too much. He issued an ultimatum to both
sides. Shelling into the exclusion zones had to stop. Weapons taken from
collection points were to be returned. All heavy weapons in the exclusion
zone around Sarajevo were to be withdrawn. Air strikes would follow if the
demands were not met.

But new NATO air raids following the murderous bombardment of the 'safe haven' of Tuzla on 25 May 1995 (71 dead, 150 wounded) did not bring the Serbs to their knees. On the contrary, in June 1995 the position of the Blue Helmets on the ground became so endangered that soon the UN had to negotiate with the Bosnian Serbs over the release of more than 370 Blue Helmets – many of whom were used as a 'human shield' taken hostage by the BSA (Bosnian Serb Army). The international community was forced to redefine the UNPROFOR mandate to protect the peace force: it was decided to regroup the forces – concentrating them around Sarajevo – thereby raising expectations of a UN willingness to give up some of the most vulnerable 'safe areas' in the process. This implicit concession by the UN was a signal that did not go unnoticed by the Bosnian-Serb leaders.

THE SHAME OF SREBRENICA

Srebrenica was the first Bosnian enclave to be protected by the UN in 1993. The decision was taken by French General Morillon during his visit in March. This personal and televised initiative forced the UN Security Council, under pressure from international public opinion, to vote Resolution 819 in April 1993 declaring Srebrenica a 'safe area'. The UN found itself in an ambiguous situation. Right from the start, the Secretary-General himself did not believe it was possible to ensure protection. At the beginning of July 1995, two years later, sacrificed by the West, Srebrenica was to be the first security zone to disappear (a few weeks later, the enclave of Zepa had a similar fate). 'Safe areas' had turned into 'killing zones'.

Some 40,000 civilians lived in this small, 200 square-kilometre enclave under inhumane conditions. Most were women, children and elderly. The strict blockade by the BSA had made a ghetto of Srebrenica, making it completely dependent on international aid for survival. Here the right of the aggressor prevailed. For more than two years, the presence of UN soldiers had frozen the military situation of this demilitarised enclave. Although they prevented any massacres, the Blue Helmets also played into the hands of the Serbs. The Bosnians found themselves forced to embrace the role of victims unable to do anything but await their fate. The trickle of humanitarian convoys that were allowed through to alleviate severe shortages brought only the bare minimum: medicines, flour, beans, oil, and sometimes shoes and clothing. Under the permanent threat of the Serb artillery and snipers, the beleaguered inhabitants (80 per cent of whom were refugees) were essentially on their own, their survival wholly dependent on the international community.

At the beginning of 1995, the Serb leadership, encouraged by ever louder calls to withdraw UNPROFOR from Bosnia, decided to impose a humanitarian blockade on the Bosnian enclaves by putting intolerable limitations

Plate 8.2
A Sarajevo bridge. The war has inflicted enormous costs on civilians, material goods and infrastructure.
Photograph: Klaas Fopma

on all food and drugs convoys. This also meant that humanitarian organisations and UNPROFOR were no longer able to replace staff in the enclaves or renew their own supplies. 'They are not hostages. If they want, they may leave,' was the cynical comment of the Bosnian Serb Vice-President Nicolas Koljevic in Pale in May 1995. Determined to follow a policy of 'an eye for an eye, a tooth for a tooth', Radovan Karadzic declared: 'If the Muslims continue to use the enclaves, the UN Safe Havens, to attack us, we shall cut off all humanitarian aid, and they may count on counter-attacks.' This message could hardly be misunderstood. The lack of international response to these threats eventually led to the massacre of the inhabitants of Srebrenica. When General Mladic opened his attack on the enclave on 6 July, he knew already the West would not react.

On 10 July, following four days of heavy shelling, Serb troops were on the verge of overrunning the town. Srebrenica's hospital was filled to capacity. More than 80 wounded lay in the corridors while shells rained down. Surgical operations were performed one after another. MSF's medical teams requested assistance from Dutch UNPROFOR doctors based in Potocari, in the northern sector of the enclave. The Dutch refused, maintaining that UN medical teams were only on stand-by to treat possible casualties among their own troops. This rejection of medical aid to wounded civilians was only another cold confirmation of the immense helplessness

of the besieged Bosnians and the cynicism of western non-interference. On 11 July, totally numbed by the shelling, the population fled to the UNPROFOR base at Potocari as their last hope for protection under international auspices.

Serb troops shot at the long lines of refugees and quickly took control of the enclave. They disarmed the Blue Helmets, who offered no resistance. UNPROFOR dispatched a single symbolic air-strike, but the situation had already proved hopeless. Srebrenica was no longer a safe haven and the civilian deportations began the next day. Men were ruthlessly separated from the women and the children, and hauled off to an unknown destination. Most would never be seen again. The 'evacuation' of the civilians had been carefully planned, with dozens of buses commandeered to transport the remaining population to Kladanj, on Bosnian territory. Within two days, Srebrenica had become a ghost town. Over 30,000 civilians were deported. Others tried to flee into the woods, but were mercilessly hunted down. From what is now clear, thousands of civilians, especially men and wounded people, were systematically butchered. According to the International Committee of the Red Cross, up to 8,000 people remain unaccounted for, including 3,000 who were last seen in Bosnian Serb hands and 5,000 who were reported to have attempted to reach central Bosnia on foot through the woods. The UN allowed these killings to happen without

Plate 8.4
Graveyards, such as this mainly Muslim cemetery, are numerous and recent in today's Sarajevo. Photograph: Jan van Veen

reacting. The Blue Helmets had become mere pawns in the ethnic cleansing conducted by the Serb militias. Broadly reported by the media, this new massacre was without a doubt one of the most repulsive to have occurred since the beginning of the war in Bosnia.

Countless UN Security Council resolutions have mandated the use of force by UNPROFOR, but only to ensure the safety of its own troops and the delivery of humanitarian aid. A clear mandate to protect civilian populations was never given because of an evident lack of political will by western governments. This allowed the massacre in Srebrenica to take place. The population could have been protected by the use of air power to stop the Bosnian Serb offensive. This was requested and approved by some UN officials. This option, however, was blocked at the highest level. These are the people who should share the responsibility for what was allowed to occur.

What happened to the populations of Srebrenica and Zepa posed a particular dilemma for the aid agencies. Should they have used their resources to remove people to safety? They were reluctant to do so for fear of abetting ethnic cleansing. The preferred option was to bring the aid to the populations in danger. This opened up agencies to Serb accusations that they were being used by the Bosnian government to maintain a hold on territory it was unable to defend. Another option was to use advocacy in an attempt to have some impact on political decisions through the lever of public opinion. During the Srebrenica ordeal, however, the advocacy campaign proved fruitless. Even with the horrors directly recounted by expatriates in the enclave, the decision-makers were not sufficiently convinced. Nor was there sufficient public or internal pressure for them to take the moral decision to intervene and put a stop to the disaster.

THE TURN OF THE TIDE

After the fall of the 'safe areas' of Srebrenica and Zepa in July 1995, there was reason to fear a BSA attack on Gorazde. However, the events of August 1995 in Krajina exposed Bosnian Serb military weaknesses. A warning of things to come had been given when the Croatian assault on Western Slavonia in May 1995, another part of the self-proclaimed Republic of Serb Krajina, was successfully concluded in a matter of hours. The rout of the Krajina Serb forces was complete, and the exodus of around 200,000 Serb refugees began. Ethnic cleansing had clearly become a strategy adopted by all parties to the conflict. Pressure was mounting against the Bosnian Serbs from all sides. Serbia's President Milosevic was anxious to have the crippling economic sanctions against his country lifted. Serbia paid the wages of the BSA officers. It also allowed military supplies to cross the border, despite the sanctions imposed on the Bosnian Serbs. These represented powerful levers with which to influence the Pale regime. Moreover, a Rapid Reaction Force was set up by the allies. Although this took time to deploy, its very existence and fire power, coupled with air-strikes, reminded the Bosnian Serbs of the disparity between their military capability and that of the UN and NATO.

Signs of a change in UN strategy had become apparent. The French-preferred option of having UN troops on the ground to do the peace-keeping was being abandoned for the US-preferred option of NATO air-strikes to enforce peace. The retreat of the UNPROFOR Britbat (British Battalion) in August 1995 from Gorazde symbolically signalled this change. Hostage-taking by the Bosnian Serbs would no longer block a forceful response. Following the shelling of Sarajevo on 28 August 1995 in which 37 people died, NATO's Deliberate Force air-strikes reacted by attacking BSA's military infrastructure. Ninety targets in 23 areas were hit. After two weeks and 850 bombing missions, they were beginning to run out of targets. The BSA, having defied the UN for more than three years, had in effect been knocked out.

In a show of Federation military solidarity, the Bosnian government forces and Bosnian Croats took advantage of the situation to launch a joint offensive in Western and Central Bosnia. These actions relieved the Bosnian Serbs of 20 per cent of their hitherto controlled territory and reunited the Bihac pocket with the rest of the Federation territory. The situation developed rapidly. The Bosnian Serbs agreed to be represented in peace talks by the rump Yugoslavia and began moving their heavy weapons from the Sarajevo area. An agreement on basic principles for a peace deal was made at Geneva, leading to a 60-day cease-fire that came into effect on 12 October 1995. Although some fighting continued after this, the new territorial lines that had been drawn were to serve as the basis for the Dayton peace talks the following month. Force of arms had at last brought the war in Bosnia to some kind of resolution.

THE DAYTON ACCORDS

On 14 December 1995 at the Elysée Palace in Paris, Alija Izetbegovic, Franjo Tudjman and Slobodan Milosevic, the presidents respectively of Bosnia-Hercegovina, Croatia and Serbia, signed the Peace Agreement on Bosnia-Hercegovina. This was the result of the peace talks at Dayton, Ohio, in November, which itself was the outcome of a major American-led diplomatic round of negotiations. The agreement provided for a unitary state of Bosnia-Hercegovina, consisting of two entities, the Bosnian–Croat Federation controlling 51 per cent of the territory, and the Serb Republic of Bosnia controlling the remainder. Each entity was given the right to have special relations with neighbouring countries (reflecting the close ties the Bosnian Croats and Bosnian Serbs had with Zagreb and Belgrade respectively) conditional upon these relations not conflicting with the sovereignty and territorial integrity of the state of Bosnia. A central government was

Map 8.2 Bosnia-Hercegovina: The areas of the Muslim-Croat Federation and the Bosnian Serb Republic are shown.

to be elected, with a collective presidency and a parliament, to be based in Sarajevo. Refugees were to be allowed back to their homes, and to receive compensation for damage and loss to possessions and property. Freedom of movement was to be guaranteed by both entities. The Dayton accords also charged a new NATO-led Implementation Force (IFOR) with creating a zone of separation between the warring armies, and to oversee the removal of weapons and minefields. They loosely tasked IFOR with aiding the resettlement of refugees and the creation of safe conditions in which free elections could be held, including the protection of civilians from violence.

The Dayton peace accord is the first step towards a lasting peace since the beginning of the conflict. However, the price of this peace is the division of Bosnia along ethnic lines. A political decision was finally made to accept the very option that had been refused since the beginning of the war and that had been the main drive for all the atrocities committed. The major flaw in this agreement is that, among the political leaders who signed or authorised this peace, are the very people responsible for the outbreak of the war and its justification. In this context, is it possible to envisage a fair and solid peace process?

A POPULATION STILL IN DANGER

By mid-1996 the benefits of peace were beginning to be seen in many parts of the country. However, the true test for the establishment of a lasting peace will be the parties' compliance with the human rights provisions of the Dayton agreement and other international instruments.

If Bosnia-Hercegovina is to become a united country again, freedom of movement and safe return of refugees to their homes are the means of implementing it. There is no indication that such movements of people are taking place on any scale large enough to suggest that the process is under way. There is no overall trust between the different communities. The result was thus predictable and understandable. The sentiments behind it are felt throughout the area. The return of refugees and displaced persons is gradually increasing. They are limited, however, to people heading back to areas under the control of authorities who share their nationality. In Sarajevo, when the transition to Federation control was completed on 19 March 1996, 60,000 to 70,000 Serbs had left the suburbs of Serb-held towns in eastern Bosnia. Both the Federation and Bosnian Serb leaders were blamed for exerting pressure on Serbs to abandon their homes. Since then, some 15,000 Bosnians and Croats have moved back to this area. But Serb families who left for the Serb Republic have yet to go back to Federation areas. Only some 8,000 Serbs have remained in the city. The division of Bosnia along ethnic lines, as agreed in Dayton, has continued. The initial UN High Commissioner for Refugees' (UNHCR) assumption was that up to 500,000 internally displaced persons would return to Bosnia, while 370,000 refugees from surrounding countries and elsewhere would repatriate during 1996. UNHCR now considers these figures optimistic. Much will depend on security conditions, funding for large-scale reconstruction, and demining, given that some 5 million landmines are believed to lie scattered around the country.

The need for humanitarian assistance for the people of Bosnia and Hercegovina remains urgent. There are still more than one million internally displaced persons in the country, many of whom have lost their homes. The UN requested a total of $823.2 million for planned projects in 1996, which include reconciliation and confidence-building measures, demining and capacity building activities. The World Bank estimates that it will cost more than 5 billion dollars to restore necessary services, core industries and communications. Yet it seems that Dayton has failed to impress the donors. Many are reluctant to finance a programme of reconstruction in an atmosphere of political uncertainty. Few have hidden their fears that Bosnia could

A trip to Zepa

The following is an extract from the diary of Michael Toole while he was accompanying a joint UNHCR/MSF convoy to the eastern Bosnian enclave of Zepa in March 1993.

10 March 1993

We arrive at Rogatica around 3.00 p.m. and park at the Serbian police checkpoint. Our driver and interpreter, a 21-year-old Serbian engineering student from Belgrade, is immediately regaled with stories that the Zepa Muslims have become cannibals and that he will most probably not make it out alive. It is too late to start checking the contents of the convoy so the French and Belgian UNPROFOR escort set up camp in an abandoned brickworks on the edge of town; we spend the night in a private home next door. Rogatica shows ample evidence of war-induced destruction. One corner of the town has been heavily shelled; burned-out remains of houses are scattered throughout the town in a random fashion. These are the homes of Muslims who had once lived peacefully in this town whose pre-war population was 12,000.

11 March

The next morning, the air is frigid. The convoy reassembles at the check-point where each truck is systematically unloaded and every box and sack counted and its contents checked by the police. The unloading is done by a group of 15 male Muslim prisoners; most of them are quite old, a few are adolescents, and none appears to be of combat age. If the number of boxes in a truck does not correspond with the convoy manifest, the excess is retained by the police. Paranoia reigns all day long. The police commander demands that half of the 89 cartons of medical supplies provided by MSF-Belgium be given to the Rogatica Hospital. We had visited the small hospital and observed some shortages in medical supplies; however, it had received a donation of Red Cross drugs the week before. While the hospital director is modest in his request for a share of the Zepa supplies, the police commander is insistent. The wrangling continues for hours in a threatening environment until a compromise is reached. Fifteen of the 89 cartons are left behind. Around 2.00 p.m., the checks are completed, and we are allowed to continue. The convoy arrives an hour later at the last Serbian checkpoint; soon after, we descend into the Zepa valley at dusk.

12 March

At dawn, we have our first look at Zepa by daylight through a heavy snow-fall. The results of months of Serbian shelling are evident everywhere; almost no building has been spared. Of the 33 villages before the war, seven have been destroyed completely. The mayor, Dr Benjamin Kulovac – a fit-looking, bearded man in his early thirties – takes us through the hospital. It is really a small dispensary, with two rooms converted into wards; in the dim yellow lamplight, we can barely make out the eight in-patients, six of whom are recovering from war injuries. Every window is broken and has been replaced with crude plastic sheeting. A huge hole beneath one of the windows has been covered with wooden planks; a shell entered there, passed through the small ward, across the corridor, and exited through the only bathroom situated on the other side of the building.

We visit several houses and speak with many of the refugees who streamed into the valley last summer; in one basement room, four women, a teenage girl and an old man lie under blankets on the floor. The old man has vacant eyes and does not speak; he seems to have removed himself from the reality of this room and of the events outside. Six months earlier, these six Muslims from Rogatica had been exchanged for the bodies of dead Serb soldiers. Prior to the exchange, they had endured the most awful torments in a detention centre in the town; the young girl, we are told later by Benjamin, had been raped. The girl's mother becomes hysterical when we start to leave; she pleads with us to help them leave the valley. She cries that they have nothing, only these blankets and the clothes they are wearing. She wants to escape to the relative security of Tuzla, or even Gorazde. The UNPROFOR escort is eager to leave and so we reluctantly get into our vehicle and wend our way back up the mountain, feeling helpless and depressed. At the top of the ridge, I try to catch a last glimpse of Zepa, but a curtain of snow removes it from sight.

Mike Toole, Australian doctor, Bosnia

well enter another cycle of destruction. Less than 1 per cent of the money needed had been made available for use in Bosnia by the end of the first quarter of 1996.

The international community should not forget the thousands of missing persons in Bosnia and Hercegovina and Croatia who lie buried in more than 300 suspected mass graves. The link between the peace process and justice should not be underestimated. Justice should not be considered a luxury, but rather as an essential therapy. It is not a question of vengeance, but of providing essential landmarks and bearings for the victims, allowing them to rebuild and to project themselves peacefully into the future. That is why the perpetrators of crimes against humanity should be prosecuted, heard and judged. This will only be possible if the International Criminal Tribunal in The Hague (ICTY), which is charged with prosecuting those responsible for the killings, receives the means necessary to fulfil its task responsibly. It is vital, too, that all countries involved be willing to cooperate by, for example, extraditing indicted individuals from their territory. Putting an end to the immunity of killers is an essential condition for peace and the restoration of trust between the different communities.

Effective action by the tribunal is needed if justice is to prevail. Days after the fall of Srebenica, Karadzic and Mladic were formally indicted for war crimes, including crimes against humanity and genocide. Already in April 1995, both had been named as suspected war criminals and summoned to appear before the ICTY at The Hague. This is a first step. According to the Dayton accords, none of the persons indicted by the ICTY may hold political or military responsibilities. It is important this provision be applied. IFOR has been given the task of arresting indicted war criminals, although not actively to seek them out. Karadzic and Mladic regularly pass through IFOR checkpoints without any attempts to arrest them. The result is that only those who give themselves up, or who are handed over by local authorities, will come to The Hague to stand trial. The ringleaders remain at large, and by mid-1996, only eight of the 78 men indicted had been apprehended.

POTENTIAL AREAS OF CONFLICT

It is the continued prospect of war which is causing the greatest concern. Despite the cessation of hostilities, serious threats of renewed fighting exist. Among these are recurrent hostilities between the Bosnian government and their Bosnian Croat partners in the Federation. After some of the most vicious fighting of the war, the Bosnian Croats and Muslims appeared to settle their differences on 18 March 1994 with the signature of an accord on the creation of a Federation.

The violent break-up of this Federation is widely regarded as one of the most likely starting-points for renewed conflict in Bosnia. The Federation exists in name only. Cooperation is haphazard and driven purely by self-interest. There have been attempts to breach this undeclared impasse, but only as the result of outside prompting, particularly from the United States. As a prelude to the Dayton accords, the leaders of Croatia and Bosnia signed an agreement that it was hoped would breathe new life into the moribund Federation. A key point was the proposed reunification of Mostar, which had been divided along the line of the River Neretva between Muslim East and Croatian West since the Muslim–Croat war of 1993/4. The European Union administration in Mostar, established in the wake of the 1994 cease-fire, has not received the support needed to reunite the city, particularly from the Bosnian Croat authorities. The Bosnian Croats and Croatia itself are closely linked, as are their aims; Croatian economic and strategic control of the region is strong. They see no reason why they should concede anything to the Muslims in the eastern enclave of the city.

Another attempt to break the deadlock was made in February 1996, when an agreement was made to remove barricades and allow immediate freedom of movement, to be policed by a joint Muslim–Croatian force. Provision was made for those refugees who could prove permanent dwelling

rights according to the 1991 census to return to their homes. A similar pattern of small steps forward being taken, while the big problems remain unresolved, is evident in the rest of the Federation. They have agreed on the boundaries and division of offices in most cantons. Yet both parties in the Federation are reluctant to cede real power. The influence of Croatian President Tudjman remains of crucial importance to the future of the Federation. Like President Milosevic in Serbia, Tudjman seems content to play a waiting game, doing nothing to upset the status quo, at least while IFOR is on the ground.

Plate 8.5
Refugee women.
Photograph: Christian
Jungeblodt (Signum)

A further reason for concern is the Bosnian Serbs' determination to have their own state. Through the Dayton accords, they think they have achieved that aim, if not *de jure* then *de facto*. While Dayton in theory provides for the eventual reunification of Bosnia, the Bosnian Serbs clearly have no desire to accept this. However, the Bosnian Serbs are more vulnerable than they have been since the outbreak of war. Relations between Milosevic and the Pale leadership are still strained. The weight of world opinion is still against the Bosnian Serbs. There is no question that IFOR would meet any aggressive move by the BSA with force. The Bosnian Serbs may be divided politically, but their collective belief in an independent Bosnian Serb republic remains strong. Dayton has left enough areas of dispute to test that resolve.

Fault-lines in ex-Yugoslavia

The cornerstone of western European policy towards Bosnia has been the principle of containment. The Bosnian war could very easily have spread, and resulted in war throughout the Balkans. This has not happened, yet problems remain, some of them as intractable as anything that Bosnia has thrown up.

The future for Serbia – whose leadership had such high hopes in 1991 – looks rather bleak. Non-Serbs made up almost 40 per cent of Serbia's 1991 population of nearly 10.5 million. (The arrival of 500,000 refugees from Bosnia and latterly the Krajina region of Croatia has swollen the Serb population, but the refugees have not been a welcome addition economically.)

A strong government, loyal police force and army have together enabled Serbs to dominate political life and to impose their will on the minority ethnic groups. Politically, Serbia remains stable, despite the combined effect of international sanctions and the cost of the war, the inflationary effects of which have largely destroyed the middle class of businessmen and entrepreneurs (and have also led to the criminal underclass dominating business life). Milosevic and the ruling Serbian Socialist Party maintain their grip on affairs. In the late 1980s Milosevic stirred up nationalist feelings, a policy that has now given way to a return to the old communist ways. With territory being lost in Croatia and the situation in Bosnia unclear, nationalism is too dangerous a beast to ride now. Control is highly centralised: those who make, or indeed are privy to, Serbian policy may be numbered on one hand.

If that grip weakens, then Serbia may suffer the same fate as Bosnia. The principle of refusing to subordinate the desire for self-determination to the inviolability of international borders – which has been the underlying theme of the Bosnian Serb position – could also be followed in Serbia with the same destructive outcome.

KOSOVO, SANDZAK AND VOJVODINA

The southern province of Kosovo presents the most acute problem. It shares a border with Albania, and is dominated by 2 million Albanians – 90 per cent of the population. It was granted autonomy in 1974, but that status was rescinded in 1989. Albanian desire for both independence from Serbia and full republic status became too much for Belgrade. It ran counter to Milosevic's fostering of Serb nationalist feeling. Kosovo was the cradle of Serb civilisation; there had been a gradual drift of Serbs away from the region, but when Milosevic harked back to the Serbian defeat by the Turks on the battlefield of Kosovo Polje in 1389, its symbolic importance was revitalised, and the remaining Serbs in the province were reinvigorated. A problem for Milosevic now is how to put that to rest once more.

Since 1990, the Kosovo Albanians have essentially supported themselves, creating parallel structures outside the Serbian government and social system. They have set up their own schools and hospitals; having been disallowed from fielding their own parties in the 1990 elections, they have boycotted polls, preferring instead to hold their own elections for a parallel government-in-exile.

Although Serbia's police control Kosovo, and its military might deters any move towards armed revolt, the situation is clearly untenable in the long term. A solution that Serbian politicians have already looked at in the recent past is to partition Kosovo: to separate those areas that are the symbolic heart of the Serbian people – Pec (the home of the Serbian Orthodox Church), Kosovo Polje (the site of the battle in which defeat consigned Serbia to five centuries of Ottoman rule) – and allow the Albanians to take the rest and unite with Albania itself.

Whether either the Serbs or the Albanians are willing to give up parts of Kosovo is one question. Such a move would also involve population and territorial transfers, something that has found little international approval when mooted as a solution for other parts of former Yugoslavia (although the final interpretation and application of the Dayton accords may lead to change in opinion).

A factor making this a dangerous course for Serbia is the reaction of the Muslim-dominated Sandzak region of Serbia itself. The Muslim leaders here have strong ties to the SDA party in Bosnia (and share the same name). Just as the war has strengthened Serb nationalism, so too has the Muslim national identity. If Serbia were to cede parts of Kosovo, then this could foster instability in the already volatile Sandzak region.

For now, Serbia is trying to keep the lid of the Kosovo pressure-cooker screwed down firmly, a task it can only perform with flagrant disregard for human rights and which will prove increasingly difficult to manage over the coming years. 'The situation in Kosovo is very tense, but [it] is part of Serbia. President Milosevic can do what he wants there,' remarked an aid worker with considerable experience in the region.

Serbia faces a similar situation in the province of Vojvodina, on the border with Hungary. Vojvodina, like Kosovo, used to enjoy an autonomous status that was, to all intents and purposes, akin to being a republic within Yugoslavia. The majority population is Serb, with the largest minority being the Hungarians, who make up almost 17 per cent of the population.

The separatist movement in this agriculturally rich area has never been as strong as that seen in Kosovo and the Sandzak, but the Yugoslav wars have awakened nationalist feelings to the extent that many Vojvodinian Hungarians had no wish to fight in Serbia's battles. Furthermore, they see their ethnic identity being threatened by the arrival of more than 100,000 refugees from the Krajina.

EASTERN SLAVONIA

Croatia has almost achieved its aim of retaking the land lost to the Serbs in 1991, and has emerged as strong as any of the former Yugoslav republics. By the end of 1995's Operation Storm in the Krajina, Eastern Slavonia was the last remaining piece of Croatian territory still in Serb hands. In October 1995 President Tudjman threatened that this too would be taken by force; would Belgrade send in its troops to support the secessionist Serbs?

Conflict was avoided, with the US again taking the lead role in forcing negotiations, through an agreement that proposed a two-year transitional period after which the region would be reintegrated into Croatia, and during which the area would be demilitarised. A UN peace-keeping and administrative operation, UNTAES, would oversee the transition.

As elsewhere in former Yugoslavia, the main problem concerns population movements. Imprecise though present figures are, they suggest that there are 160,000 Serbs in the region, nearly half of whom are themselves refugees. Before the war there were 85,000 Croats, many of whom will wish to return. It seems unlikely that Croats and Serbs will be able to live together here, particularly when memories of war are so fresh.

The two-year transition period may also see a hardening of the attitude of Belgrade, which in the lead up to Dayton had been conciliatory. However, the final handover to Croatia probably rests on the actions of the people living there now – whether they choose to stay or go. If they choose to stay, and returning Croats are prevented from reoccupying their homes, then a military solution may reappear on the agenda.

Brcko, which sits astride the narrow Posavina Corridor that links the western and eastern areas of Serb-controlled Bosnia, is one of these. The fate of this strategically vital town was undecided at Dayton, and was left to arbitration. For the Bosnian Serbs it is crucial. If they do not maintain control, they lose their link with the industrially and agriculturally rich region of Banja Luka. The Bosnian Serbs would surely go back to war if Brcko were to be taken by the Bosnian government. The town of Gorazde in eastern Bosnia remains another object of Bosnian Serb territorial ambition. The town's tenuous standing is readily apparent. Gorazde relies upon the goodwill of the surrounding Bosnian Serbs, and more importantly the presence of IFOR, to maintain its links with the outside world. Gorazde's situation will remain perilous unless Bosnia is reunited. What is more, the continuing decline of the population, estimated at 27,000 by mid-1996, indicates that its people do not believe in the future of Gorazde.

According to one UNPROFOR senior official speaking at the end of 1995, there may be more cycles of violence over the next few years before peace truly settles down. In this context, one can only wonder whether allowing the rearming of the different parties without the perspective of a sustainable peace is not like playing with a landmine. The frustration of the Bosnian government over the partition of the country and its increasing internal radicalisation could also lead to a decision to take new military initiatives.

CONCLUSION

There is little evidence that the current peace has resolved the underlying conflicts in Bosnia; the war has divided Bosnia-Hercegovina along ethnic lines; the Dayton accords have legitimised this decision. Does this mean that the legitimate aspirations and principles of 'multi-ethnicity and democracy' are being abandoned?

These accords have certainly not ended the aspirations of the antagonists. The Bosnian government in Sarajevo still hopes that the country will unite under a central government. The Bosnian Serbs are convinced that they have won their independent state. The Bosnian Croats remain committed to becoming part of the Croatian state. The warring parties have been separated, but they have not settled their political differences. The struggle for power and control of territory remains unresolved.

IFOR has successfully interposed itself between the armies and there has been little inclination on the part of the various factions to test the limits of IFOR's tolerance. This is in marked contrast to the manner in which all sides – to varying degrees – treated its predecessor, UNPROFOR.

However, IFOR's mission is a short-term one, and is scheduled to end in December 1996. There is uncertainty about what will follow. With so little progress having been made towards political solutions by mid-1996, there is a growing belief that a smaller but militarily powerful IFOR 2 would have to remain in Bosnia after the withdrawal of the main force in order to prevent a return to war.

At this stage, one can only hope that the people of Bosnia can sooner or later return to a normal life. Nothing is solved yet. Since the Dayton accords agree on the division of Bosnia along ethnic lines, it will not be surprising if tension begins to grow, with ethnic cleansing continuing, between Muslims and Croats, Muslims and Serbs, Serbs and Croats. Quite possibly, however, there was no other solution. The risk of war in Bosnia remains high. If this is to be prevented, IFOR should remain in Bosnia after December 1996. As with the Dayton accords, however, it will be up to the international community to decide.

9

CHECHNYA
TOTAL WAR AND IMPERIAL QUAGMIRE

François Jean

•

On 11 December, 1994, Russian Federation forces intervened in Chechnya to 'restore the constitutional order . . . by all available means'. Two years later, the results have proved devastating: the main cities have been razed, most of the villages bombed, countless homes wrecked or pillaged; tens of thousands of civilians are dead and thousands of men are missing or have been killed in mop-up operations or filtration camps. In the course of months, this simple policing operation had turned into a quagmire for federal troops and a huge sacrifice for the civilian population. This conflict, which caused over 50,000 deaths and provoked the flight of hundreds of thousands of people, has been marked by flagrant and systematic violations of international humanitarian law. Deliberately targeted, the civilian population has been forced to suffer indiscriminate and overwhelming bombardments. The situation has been made even worse by the fact that humanitarian organisations have been severely curbed in their activities and prevented from bringing assistance to civilians caught up in fighting and shelling. This pitiless war, laden with portentous consequences for the future of democracy in Russia, also tragically reveals the complacence of western democracies in the face of massive human rights violations.

TWO CENTURIES OF RESISTANCE

The war in Chechnya is not simply the latest manifestation of a new world disorder, inspired by nationalism and so-called tribal or ethnic conflicts, and which has supposedly filled the vacuum of the Cold War, threatening the foundations of international stability. Rather it is the latest episode in a long history of resistance to Russian expansionism. The first chapter of this secular struggle was written at the end of the eighteenth century, when Mansur Ushurma and his mountain followers resisted the Russian advance in the Caucasus. But the dawn of the following century brought the retreat of the Ottomans after the fall of Anapa fortress, the 1801 voluntary adhesion of Georgia to Russia, and the cession by Iran of northern Azerbaijan to the Czarist Empire at the 1813 Treaty of Gulistan. The people of the

Caucasus found themselves alone, isolated in a mortal showdown against Moscow. From this point on, the history of this region would be one of persistent refusal to submit. The empire, czar after czar, Big Brother and tyrants would try incessantly to bend Chechnya's will. In 1818, General Ermolov, the Caucasus governor whose fruitless ten-year campaign's only lasting result was to have hardened the mountain peoples against any means of cruelty his successors could invent, wrote to Czar Alexander I, warning that '[Chechens] by their example of independence can inspire a rebellious spirit and love of freedom among even the most faithful subjects of the Empire'. His promise to 'find no peace while a single Chechen remains alive' finds its echo in Boris Yeltsin's recent declaration, 'They're mad dogs, they must be put down like mad dogs.' In fact, for the last two centuries, many of the peoples of this 'mountain of languages' representing the Caucasus have been annihilated, dispersed, exiled, or absorbed by colonisation that began in the latter half of the nineteenth century. As for the Chechens, long known for their capacity to resist, neither war nor repression could compel them to accept the law of the empire. In 1825, Ermolov's policy of terror provoked a general uprising, and it took over 30 years and several hundreds of thousands of soldiers to overcome Shamil, Imam

Map 9.1 Chechnya: The province of Chechnya lies in the south of Russia, bordering the now independent state of Georgia.

of Chechnya and Dagestan, in 1859. It was during this first Caucasian war, the Ghazavat, that Islam, which had made a late appearance in the region, became for the Chechens a force of resistance against the invader and a major source of civic and moral cohesion.

This merciless struggle, which drained Russian military power, contributed to her humiliating defeat in the Crimean War and discredited the Romanov dynasty, was a total war not only against the partisans of Shamil but against a whole people. By 1864, when the last rebels were eliminated, there were only 40,000 people left in Chechnya. The North Caucasus was then annexed, but never pacified. In 1877–8 a new war ended, with thousands of executions, mass deportations to Siberia, and the exodus of part of the population to the Ottoman Empire. Nevertheless, rebellion was never definitively quelled, and Russia was obliged to maintain the Caucasus under military administration until the 1917 Revolution.

During the civil war, the Chechens and the Dagestanis, stuck between the sickle of the dying autocracy and the hammer of nascent totalitarianism, fought the White Russian army of General Denikine, defender of 'one and indivisible' Russia and created an ephemeral 'Mountain Republic' before revolting against the Bolsheviks in 1920–1 to preserve their hard-won independence. The history of Chechnya from 1922 to 1943 is one of an almost uninterrupted succession of uprisings – in 1924, 1928, 1937, 1940 and 1942 – against Soviet normalisation, forced collectivism and the staggering repression of 1937. Such obstinate resistance could hardly be assuaged by the declaration of an autonomous Chechen–Ingush Republic in 1936.

On 23 February 1944, virtually all Chechens, along with their fellow Ingush, Balkar and Karachay Caucasians, were deported to Kazakhstan, on the pretext that they had collaborated with the Nazis – an absurd accusation since the German army had never reached their territory. Over a third of the 400,000 forced to make this nightmarish journey to the icy steppes of Central Asia died during their transportation into exile and an ensuing typhus epidemic. Meanwhile, their villages were razed, their cemeteries bulldozed, the Chechen–Ingush Republic liquidated and deleted from maps and books. Against all odds, the Chechens survived deportation and the death camps. Solzhenitsyn wrote of them in *The Gulag Archipelago*, 'They are a nation that refused to accept the psychology of submission . . . I never saw a Chechen seek to serve the authorities, or even to please them.'

After the death of Stalin, they did not wait for an authorisation to go home. Carrying their dead, they returned to reconstruct their villages. Yet even after their 1957 rehabilitation, and until the demise of the Soviet Union, they remained the object of constant repressive measures: the KGB made the region its favourite playing field; mosques were forbidden until 1979; and the actual administration of the Chechen–Ingush Republic, restored in 1957, remained firmly in the hands of the Russian government until 1989. In June 1989, Doku Zavgaev became the Republic's first

Chechen First Secretary of the Party. With the advent of *perestroika*, however, came political renewal in the form of the creation of several opposition parties. In November 1990, an All National Congress of the Chechen People held at Grozny adopted a declaration of sovereignty and elected an Executive Committee headed by Dzhokhar Dudayev, a Red Army general who commanded an air base in Estonia. These developments marked the beginning of a complex political process which reached its culmination with the *putsch* against Gorbachev.

A MERCURIAL INDEPENDENCE

During the first hours of the failed coup of 19 August 1991, the Executive Committee declared its hostility towards the putschists. It accused the authorities of the Chechen Republic of collusion, created a National Guard, which quickly took over Grozny's main public buildings and, on 6 September, compelled Zavgaev to flee. Simultaneously, Boris Yeltsin, in a move to reinforce his still shaky position as an alternative to Gorbachev, invited the Republics to take as much power as they could swallow. The new Chechen leader took his words literally. On 27 October, Dudayev garnered 85 per cent of the votes in a hastily arranged but enthusiastic presidential election. The new President declared Chechnya a sovereign state on 1 November 1991.

A week later, Yeltsin countered Dudayev's move by declaring a state of emergency and sending troops to Grozny in an attempt to reinstate the former leaders who, as supporters of the *putsch*, had been totally discredited. United in their outrage at the Russian President's arbitrary decision and its threat of military confrontation in Grozny, the Moscow parliament met in a special emergency session. It called for the crisis to be settled by political means. Two days later the Russian troops were escorted politely but firmly to the borders of Chechnya. The Chechens had by then established a *de facto* independent state, which has never been accepted by Russia. From 1992 to 1994, Chechnya suffered a succession of acts of intimidation and pressure – economic blockade, hostile propaganda, threats of intervention. During this period, the two sides continued negotiations but Moscow never clearly expressed its fervent desire to reintegrate the republic. Chechnya appeared to be just another casualty of the break-up of the Soviet Union.

In 1992, Chechnya established a secular constitution instituting a parliamentary form of government. By April the following year, however, conflict with the opposition drove Dudayev to dissolve parliament in favour of a provisional presidential regime. The opposition immediately denounced this as a dictatorship. Meanwhile, the economic situation grew progressively worse in the 'free zone' of Chechnya. A thriving but chaotic informal economy based on uncontrolled petrol exports, unofficial international trade

Map 9.2 Chechnya and its neighbours: Médecins Sans Frontières project sites are located at Grozny and Vedeno in Chechnya; Chasavjurt in Dagestan; and at Nazran in Ingushetia.

and privatisation of public property emerged in the absence of any coherent programme of economic reform. This social and economic upheaval, exacerbated by Moscow's economic sanctions, fed popular discontent and reinforced opposition to the Dudayev regime. By 1994, pressure against the independent Republic mounted as Moscow increased its support for the opposition in hopes of overturning Dudayev and installing a pro-Russian regime. But this strategy of indirect destabilisation proved a fiasco later that year when Umar Avtukhanov's provisional council forces failed in their 26 November attack against Grozny. The prisoners taken by Dudayev's troops included many Russian soldiers recruited by the FSK (the former-KGB and soon to become the FSB).

RUSSIAN INTERVENTION

After the failure of the 26 November operation, Moscow had the option of either resuming negotiations or seeking a military solution. By early December it was clear that Russia had chosen the latter. On 9 December, a presidential decree ordered Russian troops, who had been massing for months along Chechnya's borders, to 'use all available means to guarantee state security, lawfulness, the rights and freedoms of citizens, the guarding of public order.' Two years later, the reasons for this decision are still difficult to understand. Its immediate effect was to give a shot in the arm both to Chechen patriotism, which had seriously waned in the three years of economic hardship, and to the popularity of a president who had lost the support of the majority of the population.

Quite apart from an astonishing ignorance of history, to say nothing of flat contempt for any kind of compromise, this disastrous decision tragically reveals the degree to which the political system had disintegrated in Russia. It also suggested exceedingly fragile prospects for a democratic future. It is indeed indicative of the extent to which circles of power, prisoners of their own propaganda, had become divorced from reality as well as their own society. At the beginning of December, for example, Defence Minister Pavel Grachev boasted he would take Grozny in two hours, with a handful of paratroopers. It is clear that throughout this conflict, the majority of Russians have remained opposed to the war. In addition, Moscow's handling of the situation revealed the incoherence and utterly medieval character of decision-making in Russia. Marked by the overwhelming power of a small coterie of cronies, this process has relied on a form of shadowy politburo which, without concurrent agreement and in total disregard for legality (the upper house of the Russian parliament was not consulted, as the constitution demands), decided to use force to restore the constitutional order in Chechnya.

At the outset of the conflict on 11 December, federal forces encountered civilian demonstrations in the neighbouring republics of Ingushetia and Dagestan. The first armed clashes took place on 12 December 1994, the day Russian troops entered Chechen territory. It was immediately apparent that this war, which had been described as brief and joyous by irresponsible leaders, would be regrettably bloody and absurd for the soldiers who fought it. Despite the support of an immense armada of tanks, Russian ground troops proved reluctant to face a hostile population and the resistance of a handful of determined fighters. As a result, the Russian armed forces compensated with intensive bombardments – with disastrous political consequences.

On 27 December, Boris Yeltsin, who had slipped out of public sight since the beginning of the military operations, ostensibly for a nose operation, finally informed his compatriots, two-thirds of whom were opposed to the war, why he had committed the military to the bloodiest conflict since the invasion of Afghanistan. They had gone in, he announced, to fight the mafia. The Russian president's reducing the Chechen struggle for independence to a battle against crime was further proof of a skewed perception of reality and left the population nonplussed. The subsequent announcement of his decision to 'stop the bombings to save civilian lives' was a further sham, since Moscow's promises in this domain have been disproved systematically by the facts for the past two years. All the more so since, after the New Year's disaster, Russian troops would, by default, adopt a new doctrine of shambolic brutality.

On 31 December, after softening up the terrain with intensive artillery

Plate 9.1 Victims of Chechnya's continuing war, this family collects water at a pump in a destroyed neighbourhood of Grozny. Photograph: Russel Liebman

fire, federal troops launched an assault on Grozny. The Chechen capital, however, rapidly became a death trap for Russian tanks. From this point on, the Russian forces were content to crush pockets of resistance with a deluge of bombs and to occupy cities that had been transformed into fields of ruins by artillery fire and air-strikes. Throughout this entire conflict, Russian troops have shown utter contempt for civilian lives in their efforts to suppress revolt in the breakaway republic. Thus the inhabitants of Grozny were subjected to weeks of indiscriminate, massive bombings which, paradoxically, hit the Russian population hardest. Native Chechens had long since evacuated their families to the shelter of the homes of relatives in nearby villages.

Following intense fighting, federal forces finally took the presidential palace on 19 January 1995. But it was little more than a Pyrrhic victory. The Russians found themselves in a devastated capital which would take weeks to mop up using classic policing methods (arrest, torture and summary execution) but also the less orthodox methods of a fear-stricken, undisciplined army of occupation which had resorted to indiscriminate shooting, racketeering and looting. Once a bustling capital of 400,000, Grozny had suffered the fate of Carthage promised by Russian Vice-Prime Minister Sergei Shakhrav. Chechnya's other cities (Argun, Gudermes, Shali) were all in turn hammered to destruction by Moscow's war machine.

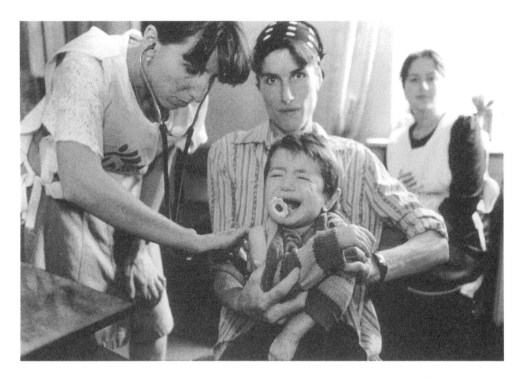

By early April 1995, Russian security forces had perfected a brutally methodical style of warfare aimed at subduing villages which had become overcrowded – particularly in the south – with displaced persons. But the outcry over the massacre of Samashki on 7–8 April prior to the 50th anniversary victory celebration over the Nazis in Moscow on 9 May persuaded Boris Yeltsin to proclaim a moratorium on military operations. This brought about a period of relative and uncharacteristic calm until the end of April. Once the celebrations were over, however, federal forces returned, bludgeoning into submission the last bastions of Chechen independence in the southern mountains. In June, Shatoy and Vedeno were both overwhelmed by tanks and air-strikes. As far as the Kremlin was concerned, the bandits had been wiped out, the war was over.

What the Russians had failed to take into account, however, was the combativeness of the Chechens. Taking up the traditions of a defensive national war against the invader, they launched a bloody raid behind Russian lines in the city of Budyennovsk. This terrorist attack, during which hundreds of hostages were taken, opened a new chapter – a political one – in this interminable war. Facing popular protest in Russia over the way federal troops were managing the problem (by indiscriminately bombing hostages and their captors with heavy artillery), Russian Prime Minister Victor Chernomyrdin opened negotiations with the Chechens.

Plate 9.2
Medical relief volunteers in Grozny treat a boy wounded while playing in the street. Photograph: Russel Liebman

A life distroyed in seconds

I work with MSF in a hospital in Vidino, a small town in the mountains of Chechnya. When we arrived here, nothing worked: there was no heating, no water and, in particular, no drugs or medical equipment. But in just a few days, with the help of the local staff, we were able to admit the first patients to the newly constructed surgical ward and to administer adequate treatment.

At first, the local population and authorities were reserved and reluctant. They simply did not understand why western doctors would go voluntarily to a war zone. Some of them, especially the Russian-speaking anaesthetist from Lithuania, thought we were spies. But after a few weeks our work earned their confidence: we repaired the water pipes, installed a generator for the operating theatre lighting, and organised stoves and firewood for the theatre and the wards.

The most impressive thing for the people of Vidino is that our services are free of charge. Before we came they had to buy their drugs on the free market, but now it no longer makes any difference whether a patient is well-off or not. The few of them who were previously able to buy advantages with a bag full of roubles have had to get used to this fact. It did not take long for the wards to fill up with patients from the area around Grozny who had been injured in the war.

Gunshot wounds are quite rare in this region because there has been no street fighting. Instead, the Russians bomb the villages south of Grozny by plane. They use fragmentation bombs, although these are forbidden by the Geneva Conventions of 1948. Many injuries are shrapnel wounds, caused by small iron pieces which penetrate the body after the explosion. Initially four new wounded were being admitted to the hospital daily, but since the beginning of March the frontline has moved closer: every day up to 20 patients need our help, in addition to the endless number of patients who would need medical treatment under peaceful circumstances.

THE BOMBING OF VIDINO

During the first weeks we experienced the Chechen war only from afar: we heard the bomb attacks; we treated the victims. But then, in the late afternoon on 4 March 1995, two helicopters approached this small mountain village from the south. Before we realised it, life in Vidino turned into a nightmare.

In a matter of seconds more than 20 rockets rained down. The helicopters disappeared again as fast as they had come and it began to get dark. Nine people, mostly women and children, had been killed and nine seriously injured. While relatives struggled to evacuate family members from the hospital, new victims arrived. The corridor was totally overcrowded; people covered in blood hoped for help, and children screamed in pain. Time was running out. We had to decide quickly whom to operate on immediately, i.e. those who would not be able to survive without intensive medical care, and who could wait some hours for treatment.

There was no question: Macha, an 18-month-old girl, had to be the first one. A bomb had landed nearby while she was in her mother's arms. It killed her mother, left her father mentally disturbed, and tore Macha's legs to shreds. She had already lost a lot of blood and was therefore deathly pale and numb with pain. James Miliken, our Irish surgeon, and Said, the paediatric surgeon from Grozny, prepared to operate. There was no choice: we had to amputate both of Macha's legs below the knees. She will be a cripple for the rest of her life although she had nothing to do with this brutal war and will most probably never understand what happened to her.

How is it possible that no one is interested in such oppression and daily cruelty against a whole population? The international community has closed their eyes and – as usual – allowed the Russians to treat the war as an internal affair. No one defends the rights of civilians. No one asks about the future of Macha and all the other victims who are completely without protection and for whom the right to life has been refused. I am sure that this ignorance and injustice will horrify me for the rest of my life.

Christine Schmitz, German nurse, Chechnya

On 23 July 1995, they signed a military agreement but failed to reach a full political agreement. Despite an unresolved situation, Chechnya enjoyed a period of relative tranquillity. The population, including refugees and displaced persons, took advantage of the break to return to their places and rebuild their homes.

But respite was short-lived. In the absence of a political solution, tension rapidly increased between Russian troops and Chechen fighters. Both sides considered that they had won the war. Tension at the beginning of October soon degenerated into full-scale combat as Moscow sought to organise a pseudo-election on 17 December, designed to legitimise Doku Zavgaev as head of the pro-Russian government based at the Russian expeditionary corps headquarters in Grozny. The truce was broken. The further devastation of Gudermes, following a week-long artillery and air bombardment in mid-December to punish an attack by a pro-independence commando, eloquently signalled Moscow's readiness to continue with its ruthless pacification methods.

A year after the beginning of the conflict, the Russian authorities still seemed to believe that force was the only way to solve the Chechen problem. Once again the Russian war machine sent tanks, planes and helicopters to launch murderous, victorious offensives against cities already reduced to rubble and villages that had already been bombed a hundred times.

Unwilling to seek a genuine political solution to the conflict, Moscow was trapped in an unending campaign to reconquer a people who, though war-weary, were increasingly radicalised by the brutality of the occupation and the savagery of repression.

By March 1996, the entire scene had a sinister quality of winter's end as Grozny found itself once again the scene of heavy fighting. Samashki, which had suffered gruesome massacres in April 1995, found itself under renewed attack and was finally destroyed in April 1996. Vedeno, invaded in June 1995, became the target of a new offensive a year later which reduced the region to ruins. Today, as in the nineteenth century, Russian troops in Chechnya control only the place where they are present, for only the time they are there. Yet their presence, combined with their willingness to indulge in sheer brutality, hardly seems to negate the overall impression of their powerlessness.

A TOTAL WAR

In Chechnya, the civilian population is the war's main victim. This has nothing to do with the excuse of collateral damages traditionally invoked by military authorities to justify the transformation of the unacceptable into the inevitable. It has more to do with the kind of war federal troops have been prepared to wage. This is a total war whose targets are not only the fighters but the entire population, young and old, men, women and children. All bear the stigma of a criminal people. And so cities and villages are subjected to massive, indiscriminate bombings which spare neither hospitals nor schools and which cost the lives of countless civilian victims. Subjugated cities are prey to mop-up operations of wholesale looting, arrests, and executions, while men disappear into filtration camps, holes of rumoured torture and terror. Worse still, it is a conflict not always considered legitimate by the Russian troops themselves. The demoralisation of soldiers, the disarray in command and control, and the lack of coordination has caused units to organise – or disorganise – themselves, with bloody consequences for the population. Spurred on by fear and vodka, many soldiers indulge in racketeering. They will shoot literally anything that moves. This has been very much the situation in the devastated capital of Grozny, whose remaining inhabitants are forced to live in a constant state of insecurity. Surrounded by bombed-out buildings and fortified barracks, people thread their way through tank patrols, the evening curfew and nocturnal barrages of gunfire. Nor has the countryside been spared. Here the civilians are hostages not of wayward soldiers but of military chiefs who blackmail them, demanding peace treaty signatures and threatening destruction of their villages with unlimited collective reprisals.

In this murderous war, humanitarian action is almost powerless. The humanitarian corridors, occasionally opened by the military to allow civilians to escape from bombed villages, often function only if one can pay. There is also no guarantee of security, least of all for men between 15 and 60, who risk arrest. Independent, impartial organisations are kept at a safe distance, where they must content themselves with waiting for refugees to arrive but remain powerless to help those in danger.

This total war has led to massive population displacement. During the first months of the conflict, the destruction of cities caused a gigantic exodus: over 100,000 people took refuge in the neighbouring republic of Ingushetia, nearly 50,000 in Dagestan, and hundreds of thousands retreated to the south of the country, to zones that had so far been spared from bombardment. But this desperate flight of nearly half the republic's population in the face of the Russian war machine did not result in the organisation of refugee camps. Most displaced persons have sought refuge with relatives and kinfolk, accommodated in family houses. Despite such solidarity, however, the situation of these populations remains exceptionally precarious.

The problem has become all the more serious given that international relief operations are constantly hampered. Despite repeated demands, United Nations agencies have not been authorised to intervene in this conflict. The rare humanitarian organisations with access to Chechnya face obstruction which is obviously politically motivated. Thus, in spring 1995, no relief convoys were allowed to move in the uncontrolled zones flooded with refugees in the south of the country. And, in spring 1996, humanitarian teams helplessly witnessed the destruction of the villages of Sernovodsk and Samashki. They were only permitted to assist the 25,000 people who had fled again to exile in Ingushetia.

Since the beginning of the conflict, this problem of access to victims has proved particularly acute in Chechnya. Humanitarian organisations which attempt to provide aid to people trapped in bombing raids and combat zones are systematically stopped and often accused of supporting the fighters or of spying. Apart from the heavy legacy of 70 years of official Soviet paranoia and the lack of familiarity with the way humanitarian organisations operate, this suspicion is also due to the confusion – well ensconced in the minds of the military authorities and lamentably evident in combat zones – between fighters and civilians.

Despite such suspicions and constraints, humanitarian action is nevertheless tolerated. But humanitarian organisations are worn out by the obstacle course presented by an infinite variety of administrative and political problems which hamper their possibilities of action. Relief workers are considered undesirable, and must never intervene in emergency situations except by patiently following the movements of the troops. And they are also a target, like any other, in this war where neither civilians nor hospitals are respected, where human rights and international humanitarian law are openly violated in an atmosphere of general indifference.

Plate 9.3
Chechnya's brutal war has caused numerous military casualties on both sides, but the main victims are civilians. Women cry on a bus in Grozny as they are evacuated.
Photograph: Russel Liebman

THE SILENT COMPLICITY OF THE WEST

For nearly two years, this pitiless war has continued with utter impunity before the crushing silence of the international community. Since the very first days of the conflict, the western countries have accepted the Kremlin's arguments at face value. These are repeated in unison, like a lesson learned by rote, maintaining that Chechnya remained an internal Russian affair, as if this could justify the unacceptable. American Secretary of State, Warren Christopher, struck a chord of approval in the first days of the conflict. 'Yeltsin has probably done what he had to do,' he announced, and added, for good measure, that the Russian authorities had 'shown as much restraint as was possible'. This cynical declaration, which gave the go-ahead in advance to the programmed crushing of a people, demonstrates that western countries were quite happy to accommodate Moscow and exchange their speeches on human rights for a brand new version of *realpolitik*.

Even silence would have been preferable to this spate of indulgent complicity, which amounted to encouraging repression while turning a blind eye to the methods employed to achieve it. Western complacency is all the more irresponsible given that, in sending its tanks into Chechnya, the Kremlin clearly demonstrated its lack of regard for its international commitments. The Code of Conduct on Politico-Military Aspects of Security,

adopted at Budapest in December 1995, just days before the Russian intervention, engaged the member States of the Organisation on Security
and Cooperation in Europe (OSCE) to refrain from using force against
minorities.

The ink on the signatures was scarcely dry before Russian troops
began bombarding what Moscow considered to be a territory of the
Russian Federation – the Chechen Republic, inhabited by a minority of
the same name. At a time when democratic countries make desperate efforts
to establish rules aimed at preventing new conflicts, this flagrant violation (with the benediction of the West) of the Code of Conduct, which
was supposed to complete the process begun in Helsinki in 1975, is ample
proof that respect for human and minority rights is no longer considered
one of the basic values of post-Cold War Europe. Russian human rights
activists, following Serguei Kovalev's example, protested against this
tragic abandonment, but the West responded with indifference and double-
talk. Sweden was the only country to refuse this generalised smugness
by declaring that Russia's means of resolving this conflict could not be
considered an internal affair. Following the American example, however,
other European countries prepared to draw a discreet shroud over the
massacres and bombardments hoping that the abscess would soon
burst.

Plate 9.4
A wounded rebel
visited by friends
during an open-air
concert at a hospital
in the Caucasus
mountains, where the
last rebel strongholds
exist. Vedeno,
southern Chechnya,
April 1995.
Photograph: Teun
Voeten

Plate 9.5
Russian tanks patrol
the streets of the
destroyed town of
Grozny, Chechnya,
April 1995.
Photograph: Teun
Voeten

As the bombings continued, and the brutality of the pacification methods spilled on to the television screens, a few weak protests were heard in the West. Rejecting any bloodshed – a ridiculous euphemism – when Grozny was already literally crushed under a deluge of fire, they decided to put off the granting of an International Monetary Fund (IMF) loan, to suspend the process of ratifying the interim agreement with the European Union, and to suspend the procedure concerning Russia's request for membership of the Council of Europe. In any case, these decisions were prompted more by a desire to create a decent interval at a time when the bombing of Grozny was still the lead story on the televised news than an actual will to influence the Kremlin to change its policy. The 6.5 billion dollar IMF loan would be granted a few weeks later without any political counterpart, the interim agreement would be signed on 17 July 1995, despite the fact that none of the European Union's conditions (the start of political negotiations, free access for humanitarian aid) had been met, and on 25 January 1996 Russia was admitted to the Council of Europe – the democratic conscience of Europe – without ever having manifested any real intent to respect European demands regarding human rights.

In the same manner, the decision of western leaders to participate in Moscow's 9 May commemoration of the 50th anniversary of victory over the Nazis, and to avoid public evocation of the subject of Chechnya, was a clear signal to the Russians that the massacres of civilians had been swept under the rug. They were now free to continue to let terror reign within their borders. Since then, western countries seem to have banished Chechnya from their list of preoccupations and have become used to the bloody rampages of the leader of the Kremlin. And this ostrich policy is accompanied by important financial support (in the form of a new 11 billion dollar IMF loan in 1996) and by eternally renewed confidence in a leader whose promises are, day after day, repudiated by the facts.

The complacency of the West seems to have been based on the intent not to humiliate a Russia unwilling to digest the fact that it is no longer a great power. At a time when Moscow deploys its immense war machine to crush a little people a million strong, it would behove the Kremlin to reassure its neighbours, particularly the countries of the near abroad, as to its intentions. On the contrary, however, it is the western countries who have rushed to reassure Russia. They have presented her with a *carte blanche* to pursue a colonial war, even at the price of massive human rights violations. This paradoxical attitude, unfortunately characteristic of relations with Moscow, may well have the reverse effect. It may lead Russians to believe that they may, with all impunity, follow the old line of imperial logic and return to their former methods. Obsessed by a false conception of stability and vulnerable to Moscow's arguments about the Islamic threat and the criminal character of the Chechens, the West seems ready to sacrifice its professed principles in the name of a short-sighted *realpolitik*.

This position is not only morally shocking, it is politically dangerous, primarily because of the imperial revival of Russia, which, using the pretext to reinstate order and keep the peace, seeks to reconstitute an exclusive sphere of influence. This can only be a source of disquiet for her neighbours, recently freed from the Soviet juggernaut. The unilateral character of initiatives by the Kremlin, in violation of international commitments, is a challenge to new security structures such as the OSCE. It also carries with it the germ of a return to 'bloc' logic, which is far from the best assurance of stability and balance in Europe. On another level, Moscow also seems incapable of providing real political solutions to conflicts, or real stability within its own territory. Threats of disintegration in the ex-USSR are certainly real, but they are not necessarily related to Chechen aspirations for independence. They have more to do with the negligence of an irresponsible power that knows only the language of force and is incapable of replacing totalitarian order with a society of organised stability.

Despite such disquieting tendencies, western countries obstinately maintain a benevolent attitude towards Moscow. This tragic blindness amounts to ignoring the mass of public opinion, which deplores the war, and encouraging the slow swing of Russia back towards authoritarianism. When the president violates the constitution by using force without consulting parliament; when the state is subject to the whims of unpredictable individuals who are out of touch with society; when civil power progressively loses control over the military; when various once-sinister security services begin to sneak back into power; when the press, once again under pressure, is obliged to disseminate official propaganda; when the rule of law is defied or ignored; when all parties, with the exception of Vladimir Zhirinovski's extreme rightists, denounce the Kremlin's irresponsibility, it is difficult to pretend that Boris Yeltsin represents the best hope for the democratisation of Russia.

And yet democratic countries continue to support the man who has dragged Russia into the bloodiest and most absurd conflict since Afghanistan. It is true that, despite the return to an autocratic and neo-imperial conception of power, the elective process has become established in Russia. As the date of the presidential election approached in late spring, Boris Yeltsin, for once taking public opinion into account, committed himself to negotiations in May 1996. But, beyond a fragile cease-fire and equally weak campaign promises, it is by no means evident that the Kremlin finally understands there can be no military solution to the conflict. Rather than locking themselves into unconditional support for Boris Yeltsin, western countries would do better to expend their efforts in promoting the quest for a political solution. Confronted with powers that use force against their own population, easy complacency is always more dangerous than reasonable firmness. Given the massive violations of human rights and of humanitarian law in Chechnya, democratic countries would do well to remind Russia of the principles they profess to uphold.

10

RWANDA TWO YEARS AFTER
AN ONGOING HUMANITARIAN CRISIS

Vincent Faber

•

The genocide of a million Rwandan Tutsis and the political massacre of moderate Hutus from April to July 1994 has left an indelible stain upon the annals of history. These dark times have presented in stark relief the human capacity for extreme and calculated brutality. Hutu and Tutsi leaders have demonstrated patent disregard for the rights of civilians, victimising and using them in their struggle for power – and they continue to do so. The international community has made a mockery of its professed vision of a 'new world order' by following inaction during the crisis with irresponsible and inadequate measures in the aftermath. Caught between the warring parties and the international community, humanitarian organisations have faced tremendous obstacles and moral dilemmas in their efforts to help civilians during the crisis, and two years later the circumstances have proved slow to change.

REASONS FOR ENTRENCHMENT

At the beginning of July 1994, the extremist Hutu leaders who had planned and led the genocide in Rwanda changed their brutal strategy. Forced to flee the country, they perceived, with a sense of anticipation that in other circumstances might have inspired admiration, the strength and protection millions of Rwandans could provide by accompanying them in their exodus. They then redirected their power to compel their Hutu compatriots to retreat from Rwanda with them. In a matter of days, 2 million Hutus – many of them virtual hostages – crossed the border into Zaire, Tanzania and Burundi. Today, there are still around 800,000 Hutus in the mega-camps of Goma, 350,000 in the Bukavu and Uvira region, 500,000 in the camps of north-western Tanzania, and 90,000 in Burundi. Over the months, these camps have become small cities, with businesses, bars, cinemas and hotels, and their administrative structures have been developed and are dominated by former Rwandan authorities. The camps hang like a veritable sword of Damocles over the Rwandan government and have fuelled increasing instability in Rwanda as well as in their host countries.

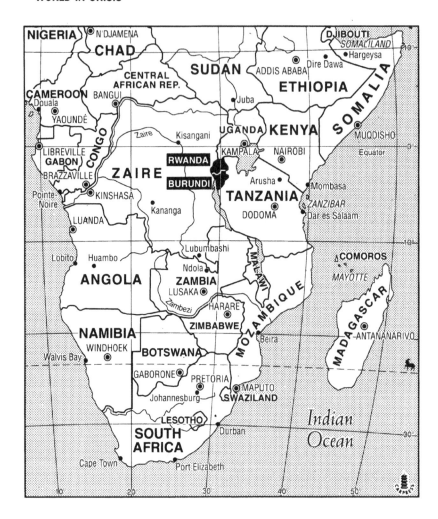

Map 10.1 Central
and Southern Africa:
Rwanda and Burundi
are situated between
the countries of Zaire
(to the west) and
Tanzania (to the
east).

As the Hutus consolidated power in the camps, the new government in
Kigali, dominated by the Rwandan Patriotic Front (FPR), attempted to
revive the country. Rwanda's moral core, its morale and its infrastructure
had been devastated by its former government. The new government had
to control the understandable desire for revenge – often expressed in unbri-
dled hatred for the Hutus – profoundly felt not only by the surviving Tutsis
but also by those who had returned from long exile abroad. Moreover, it
had to tend to the needs of a population that had survived profound psychological trauma and remained in a state of immense confusion. Structurally,
everything had to be rebuilt: banks, schools, hospitals and dispensaries,
public transportation, administration, the courts of law. Human resources,
including medical personnel, teachers, police, magistrates and engineers,
were woefully lacking.

International financial aid, which might have compensated in some way for the international community's lack of response to the genocide, was distributed in driblets. Some countries – France, for example – held stubbornly to their support of the murderous Habyarimana regime, refused to free necessary funds and continued to ostracise the country's new rulers. Duplicitous western countries deliberately refused the Rwandan government the means to carry out social and structural reconstruction – not to mention reconciliation, which it would be naive and presumptuous to mention. They insisted instead on wielding some power themselves, hypocritically flaunting their moral and democratic virtues, recently quite absent in the face of genocide, and pointing out the faults of the Kigali regime.

The new Rwandan government has not only failed to defend human rights, it has overseen outright violations. In Rwanda's western provinces, government repression has increased in response to infiltration and provocation from refugee camps just over the border. Over 73,000 people are still imprisoned in cruel and unacceptable conditions. They await trial by a justice system which is still new and incapable of judging those responsible for genocide. Recurrent, often unsanctioned massacres by the army add to the deplorable situation. The worst of these was in Kibeho in April 1995. Not only do these circumstances deter refugees from returning, but the Rwandan government has been extremely reluctant to repatriate refugees and has shown a dismissive attitude towards those who have returned. Clearly all this places humanitarian organisations in the difficult position of maintaining a balancing act, compelled to comprehend the reasons for the repressive actions of the Kigali regime, while at the same time uncategorically denouncing them. Humanitarian organisations must maintain reasonably good relations with the Rwandan authorities in order to continue programmes of assistance to populations in distress.

The international community has contributed to the failure to resolve the Rwandan crisis. Bogged down by its own contradictions and dogged by conscious or subconscious guilt for its lack of action during those dreadful days of genocide, the international community has to date been incapable of supplying anything more than a 'compassion mechanism', to use Rony Brauman's expression. During this entire crisis, the international community has demonstrated only duplicity and cynicism. The highest source of this discourse, the United Nations Security Council, has passed resolution after resolution over the past two years without ever providing the means to implement them. For example, the UN sent 2,500 Blue Helmets to pre-war Rwanda to guarantee the Arusha accords and, *ipso facto*, to protect the civilian population (Resolutions 872, 891, 893 and 909). At the same time, however, it refused to authorise protective measures to be taken and, worse, reduced the number of Blue Helmets to a tenth of their strength at the height of the violence. It roundly condemned the genocide and its instigators (Resolutions 912, 918, 925 and 935), all the while

Number 17

Six brick houses on the outskirts of Butare, reached by a hard clay track. This was Kabutare hospital, where MSF-Belgium had asked me to fill in for someone.

My Belgian colleagues had handled the Kibeho emergency, during which they had to treat hundreds of wounded people in the space of a few days. I arrived a month later, and many of the survivors were still at the hospital. They were in 40 rudimentary metal beds in the long, dark room. That morning, I was given a book containing 40 numbers, each followed by case notes.

Number 17: 'Hasn't eaten; brings up everything he is given.' Here we were, in an adult ward, and Number 17 was hiding under his blankets: he was only 12. Gradually, his neighbours let me know that he had no family left.

Since receiving a bullet in the head, he had not spoken to anyone, or eaten anything. He was a living skeleton, with match-stick arms. He had made himself invisible in the turmoil of Kibeho; now he was doing the same in hospital.

Helped by his roommates, I was putting the puzzle back together. Step 1: I lifted him out of his bed. My arm around his skinny shoulders, we walked towards the sunlight. Sitting on the steps leading down to the water tank, I sang the few words I knew of a song by Lokua Kanza: 'O shi ku n'Afrika . . .'. He clung to my side, silent, robotic. From there, we went and filched a couple of bananas from paediatrics. He devoured them, and had to stop himself choking.

Over the next few weeks, I would come back to see him every day after my work with the mobile teams. He was still hiding under the covers, but at the sound of my voice, he would throw them back and welcome me with a burst of laughter. He was gradually gaining confidence and, with it, more of the nurses' attention. But, I thought, how far should I go? My mission in Butare was due to end in a month. Should I let him get attached to me, knowing that I would leave him and he might feel abandoned – again?

Encouraged by his faint progress, his roommates and the Rwandan nurses talked to him more often in Kinyarwanda, with stubborn kindness, trying to get through to him. My questions in French were eventually answered with a nod or a shake of the head. Little by little, Number 17 took to a normal diet of rice and beans. Having become something of a character, he began to receive small treats from my colleagues in paediatrics. The other children played in the hospital corridors, laughed and ran races on their crutches, but Number 17 watched in silence.

The last day of my mission finally arrived, and my very last visit to the hospital. On arrival, I detected conspiratorial smirks on my Rwandan colleagues' faces. As I reached him, Number 17 frowned, and with a determined expression, pretending not to notice the look on my face, said his name.

Roswitha Blassel-Koch, Swiss nurse, MSF, Rwanda

refusing to arrest them or have them arrested. It continues to call for the constitution of an international tribunal to judge crimes against humanity committed during the war (Resolutions 935, 955, 977 and 978), but it has failed to provide the financial means to establish the court. It expresses the wish for the safe return of refugees and pretends to guarantee their security (Resolutions 965, 997, 1029 and 1050); yet the task is nearly impossible because the Security Council has authorised a ridiculously inadequate number of UN observers to provide that protection. Further, it has failed to dismantle the network of intimidators operating within the camps in Zaire and Tanzania.

Certain member countries of the UN, technically pledged to respect its decisions, have gone beyond a simple lack of cooperation. Mobutu's Zaire, for example, still maintains a policy of active complicity with the former Rwandan regime. They are allowed to prepare their war of reconquest and are provided with logistical facilities. Mobutu's armies help by undermining the feeble attempts by the Zairean government to prevent economic activity in the camps and to encourage the repatriation of refugees. Mobutu has given the Rwandan Army Forces (FAR), the brutal forces of the deposed regime, free rein in the Masisi, which they are attempting to colonise, to create for themselves what Uganda has been for the FPR for 30 years.

The long-term nature of the camps lends structure and normality to the lives of the refugees and renewed power to the authors of genocide; combined with the hardening of the new Rwandan regime in response to the threat brewing at its borders, plus the weakness and indecision of the international community, the crisis has become entrenched. Meanwhile, human rights violations increase, becoming almost commonplace, as the Rwandans live in a state of perpetual moral and physical distress; hatred spreads throughout bordering regions, and the refugees live in a state of unending exile. These dark clouds slowly gather around the Great Lakes, promising new turmoil in the future. In this context, humanitarian organisations, their presence indispensable in a situation of such obvious need, are faced with increasing doubts as to their role and how they are to accomplish their tasks.

REFUGEE CAMPS AND HUMANITARIAN DILEMMAS

The cholera epidemic, which followed close on the heels of the arrival of millions of refugees in Goma in July 1994, cost 50,000 lives. The international community responded with unprecedented levels of humanitarian aid to the camps. The action was the subject of great controversy, since it stood in stark contrast to international passivity in the face of genocide only weeks before. Without going into the details of this event, already examined in several works, it is nevertheless clear that the massive injection of aid gave rise to the worst excesses, including the perversion of

The forgotten war of Masisi

Lying some 50 kilometres away from Goma, Masisi was long considered the 'Switzerland of Africa' for its green valleys and rich pastures and the strength of its market, among the most prosperous in all Zaire. Today, Masisi is devastated and deserted. The rare traveller sees only a succession of burned villages, and those who could not flee live in constant terror, under the combined threat of militia machetes and the pangs of starvation.

Masisi was originally inhabited mainly by the Nande, Nyanga, Tembo and Hunde tribes, but after the Berlin Conference decided colonial frontiers in 1885, some Hutus and Tutsis found themselves on the Zairean side of the border with Rwanda. Throughout the twentieth century, Masisi has seen a constant influx of Banyarwandas ('those from Rwanda'). Whether Hutus or Tutsis, these people have come to Masisi to seek refuge each time starvation or massacres cast a deadly shadow over their homeland. In 1994, an estimated 150,000 Hundes lived in Masisi along with 450,000 Hutu and 50,000 Tutsis out of a population of approximately 800,000.

As early as the 1970s, the first ethnic tensions appeared. The Banyarwandas, Hutus and Tutsis who had grouped together recognised that their strength lay in their numbers and at this time demanded a share of regional and customary power, then held entirely by the Hundes. Given their deep roots in the area, often several generations old, they also requested full Zairean citizenship. After their requests were repeatedly ignored, the Banyarwandas took up arms. Masisi was then precipitated, in 1993, into its first war, which killed more than 20,000 people. The outside world was entirely indifferent. ICRC [International Committee of the Red Cross], Oxfam and MSF teams, already present in Goma, tried, despite their insufficient means, to provide care, food and assistance to the nearly 250,000 displaced by the war.

In early 1994, the warring parties made a preliminary agreement and requested the Zairean government's assistance in strengthening the peace. The fighting came to an end, and some returned to their homes, but this seeming return to normal life would not last. Kinshasa foolishly expressed disdain for the agreement, and suddenly the region was overtaken by a massive influx of a million Hutus fleeing Rwanda in July 1994.

The conflict flared up again in the autumn of 1995, more cruel, more barbaric than ever. Refugee camp leaders joined the renewed fighting, seeing an opportunity to create the 'Hutuland' they had dreamed of and even from there making their conquest of Rwanda.

Today, massacres endlessly follow massacres, villages are in flames, and the whole area is ruled by a vicious cycle of reprisals and counter-reprisals. In only a few months, the death toll reached over 10,000, and over 300,000 people – from all ethnic backgrounds – are, once again, displaced. The Tutsis, once allied in the same Banyarwanda brotherhood with their Hutu fellows, now find themselves targeted from all sides. Only one or two thousand survive, left to their deadly fate.

MSF and the ICRC are the only international agencies providing assistance in the area. But our volunteers must attempt to overcome tremendous obstacles: the unpredictability of the conflict; the dubious nature of the chains of command (if they exist at all) of the fighting factions; the two hundred or so Zairean 'peace-keeping' soldiers who find their only sustenance by taking it from the local populations; and, finally, the state of decay of the local infrastructure. We have had to postpone or cancel many operations due to security risks, and sometimes find ourselves unwittingly supporting ethnic cleansing efforts by recommending Tutsi populations evacuate certain areas.

Despite its enormous logistical and financial resources in the area, the international community remains indifferent and inactive. Zairean authorities have shown scandalous passivity. Our presence in Masisi and in Goma is more necessary than ever, not only because we bring essential humanitarian assistance – but we also remind the world that this barbaric and bloody conflict, ignored by all, still rages on.

Vincent Faber, former MSF country manager in Zaire

providing the former Hutu authorities (involuntarily of course) with a means of regeneration.

Médecins Sans Frontières was one of the first to point out these abuses and to denounce them loudly, but the effort was futile. UN agencies, including the United Nations High Commissioner for Refugees (UNHCR), were limited in their response because of political constraints, and important humanitarian organisations showed no interest in the situation. Donor countries, determined to maintain a high level of aid in order to divert attention from earlier impotent policies, ignored the abuses. In addition to an unreceptive audience, MSF faced physical threats to its teams in the field as well. The search for an alternative policy provoked considerable debate within MSF.

On one side, there were those who advised immediate withdrawal from the camps, since compromise with the perpetrators of genocide was impossible to stomach and the continuation of aid merely reinforced a situation in which hundreds of thousands of innocent refugees were, in effect, being held hostage. But in that case, these populations would be left on their own, with inadequate access to the medical care, health assistance and food essential to their existence. Resources for the genocidal criminals would be cut off, but so too would those for the refugees, the majority of whom are innocent victims of odious manipulation by the extremists. On the other side of the debate were those who recommended a continuation of humanitarian action. The need was too extreme to ignore, they argued, even at the price of conscious, if passive, complicity with those guilty of crimes

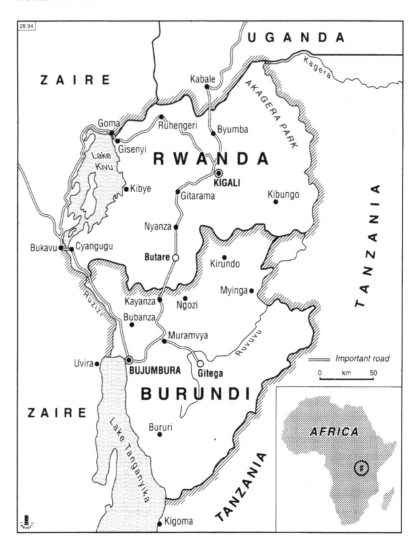

Map 10.2   Rwanda and Burundi: Médecins Sans Frontières project sites are located in Cyangugu, Kibungo, Gysenyi, Butare, Kigali and Ruhengeri in Rwanda; Ruyigi (east of Gitega), Kirundo, Makamba (south-east of Bururi), Ngozi, Karuzi (south-west of Myinga), Muzema (south of Kayanza) and Bujumbura in Burundi; and at Goma and Uvira in Zaire.

against humanity, and indirect participation in the perpetuation of the threat of future genocide.

We at MSF were faced with a cruel dilemma, which boiled down to humanitarian ethics vs. medical duty, moral responsibility vs. the Hippocratic oath. There was no good choice; there was only the lesser of two evils. Of course, we reduced our activities in the camps to the vital minimum, and we continued, often successfully, to pressure the UNHCR to set up the necessary controls to stop at least the most blatant siphoning off of aid. At the same time, the strategically sophisticated extremist leaders had understood that their structures of control would have to be more discreet if they wanted to maintain their presence and their political networks in the camps. So, in the first half of 1995, they accepted these changes: the establishment of a familial system of distribution designed to avoid misappropriation of food; exclusion of the most visible ex-FAR military forces; democratic elections to choose administrative authorities for the camps, etc. But we knew these were only cosmetic changes and that, basically, nothing had really changed. And so, once the humanitarian situation permitted, we left, all of us this time, in the autumn of 1995, having delegated our activities to other organisations. Our medical duty fulfilled, we were now at liberty to address our moral responsibility. Was it the right choice? It is hard to say at present, and we will have to wait months, perhaps years, to know for sure. One thing is certain: the extremists were delighted to see us go, relieving them of the stick-in-the-muds who prevented the massacres. And, clearly, the possibility of observing and reporting any renewed threat of harm in the camps, from these genocidal forces, has now been greatly weakened, if not practically eliminated, due to our absence.

Another dilemma we faced was what position to take regarding the difficult subject of repatriation. While we recognised that the existence of the camps was essential to the political survival of leaders of the former Rwandan regime, and that the camps themselves were the breeding ground for further genocide, we also saw that Rwanda was incapable of providing the basic guarantees necessary for a safe return of the refugees. For a long time, our policy was to support repatriation only of those who freely expressed their wish to return. Theoretically, this was perfect; but because a handful of leaders controlled the immense majority of refugees through constant intimidation and misinformation, and the few candidates for repatriation were found massacred inside the camps, it was clear no one could freely express their desire to return. For a long time, despite UNHCR measures to protect and accompany the returnees in the camps, only 500 to 1,000 a week dared to take this step. The process did speed up, however. The passage of time weakened the leaders' powers of intimidation, and, undoubtedly, the reassuring messages emanating from the UNHCR and other humanitarian organisations across the border began to gain credibility among the refugees.

But just as repatriation was on the rise, the intimidators themselves found an unexpectedly influential ally: the Rwandan government. Exactly one year to the day after the inception of genocide, in April 1995, news began filtering through to the camps about massacres of Hutus carried out by the Rwandan government forces (APR). And on 22 April stories spread about the Kibeho massacre. From one day to the next, the refugees perceived that their chief intimidators were no longer in the camps, but in Kigali. The Hutu extremist leaders no longer had to fabricate stories or use violence to intimidate the refugees. They simply spread word of the dreadful conditions in Rwandan prisons, the killings by the APR, and the absence of an effective process of social and economic reintegration for the repatriated.

Plate 10.1
Refugees in Kibeho, Rwanda, clamber to get water. Soon thereafter, many are massacred in a brutal slaughter. Photograph: Charly Brown

The stories dealt a crushing blow to the UNHCR's feeble attempts to encourage repatriation. By the summer of 1995 only about a dozen refugees chose to return each week.

The Zairean government of Kengo, pressured by increasingly discontented local populations resentful of the refugees, attempted to force the refugees to leave. To the Zairean government, the inhabitants of Kivu are an essential element in their political fight with Mobutu's forces, and therefore, when faced with numerous delays, they decided to take matters into their own hands. On 31 August 1995, Zairean armed forces rounded up immense numbers of refugees at the Mugunga camp, near Goma, and deported them *manu militari* to the nearest border crossing. Approximately 17,000 refugees were forcibly repatriated over three days. Contrary to numerous reports, the operation was carried out with a minimum of violence, as MSF volunteers witnessed. Although many saw in this albeit maladroit action a means of dissolving the Hutu leaders' influence, at MSF we were obliged to firmly condemn the act. In the face of universal disapproval, Zaire interrupted the operation; but throughout the winter it called for the dismantling of the camps with increasing insistence, impatiently demanding that the UNHCR and NGOs (non-governmental organisations) take more vigorous action in

Plate 10.2
United Nations soldiers attend to the mass graves for the victims of the Kibeho slaughter.
Photograph: Charly Brown

the matter. In February of this year, Zaire carried out a similar, though less coercive, military operation. The army encircled the camp of Kibumba – at 5 kilometres from Rwanda, the closest one to the border – so as to restrain the movements of the 200,000 refugees and to forbid all commercial activity. But again, the loose blockade that was planned as the 'administrative shut-down of Kibumba' was utterly ineffective.

Insecurity among the refugees grew, and they began to look for alternatives. They feared Zaire's so far awkward attempts to induce repatriation would be transformed into a blatant, violent round-up-and-expel operation. Is it a mere coincidence that, in autumn 1995, as Zaire expressed its impatience with increasing frequency and explicitness, the neighbouring region of Masisi flared up once again, this time in a manner that, on a smaller scale, mirrored events in Rwanda in 1994? Most likely not. The new flare-up in the Masisi involved modern arms previously unknown there, and it is likely that the Hutu extremist Interahamwe and the ex-FAR were directly involved. The Masisi would be a perfect base for a Hutu reconquest of Rwanda. Its proximity to the border and its majority-Hutu population would make it a perfect logistical platform and the ideal 'Hutuland' for those wary of staying in the camps.

RWANDA: HUMANITARIAN CONFUSION

April 1995 will remain a watershed in post-genocide Rwandan history. Within a few days, starting on 18 April, 250,000 internally displaced people, grouped in a dozen camps in Gikongoro prefecture in south-west Rwanda, were forcibly evacuated and sent to their communes of origin. In the largest of these camps, at Kibeho, which held around 100,000 refugees, the evacuation ended in an indiscriminate and widespread massacre. On 22 April, the APR fired for nearly eight hours without respite upon unarmed civilians who had been encircled for days without food, water, sanitary installations or medical facilities. All this took place without any reaction from UNAMIR (UN Assistance Mission for Rwanda), the UN contingent present, which was assigned to protect the refugees. The survivors were forcibly returned to their communes, unaccompanied by international observers. Most were stoned and harassed along the way.

MSF volunteers witnessed these events and estimated the number of victims at over 3,000, including many women, children and elderly. The international commission of inquest set up at Kigali's request several weeks after the tragedy fixed the number of victims at a little over 300. This discrepancy was likely to be deliberate, intended to free UNAMIR from blame, since they were already the object of much controversy. The UN was negotiating with Kigali at the time for UNAMIR's continued presence in Rwanda. For this same reason, the commission of inquest laid part of the responsibility for the massacre on the shoulders of the NGOs.

Burundi, complex emergency – moving up the charts

After more than two years of silence, Burundi made the limelight in December 1995. It has remained there ever since. But with the world attention focusing once again on this small, central African nation, nobody knows what song to sing. Backstage, a slow, smouldering fire is consuming the theatre. The 35-year-old post-colonial power struggle in Burundi ironically reached a peak following one of those rare peaceful political transitions in African history. The assassination of the first democratically elected president, Melchior Ndadaye, and the ensuing revenge and retaliation caused the death of tens of thousands. But with the collective suicide of neighbouring Rwanda, Burundi froze in utter panic. In the period that followed the conflict, it developed, in the eyes of international community, from a chronic nuisance to a complex emergency: Burundi had to be saved.

The initial strategy was one of containment, sustained by silence and positive diplomacy. Out of fear of creating a self-fulfilling prophecy, the deterioration of the humanitarian situation and the escalation of violence were down-played by international politicians and humanitarian agencies alike. Yet while the conflict remained contained, nothing was done to solve it. Meanwhile, human rights abuses multiplied. Massacres grew in size and number, the economy collapsed, and guerrilla pressure increased. The policy of containment had failed. The need to find a solution had become even more pressing.

Respect for humanitarian principles and those seeking to uphold them were not high on the agenda of the warring parties. Relief workers have found themselves alarmingly restricted in their efforts to intervene. Civilian wounded cannot be collected in acute conflict-zones and neither can displaced persons forced to flee their homes be provided with water or shelter. The gaps in humanitarian assistance are widening as aid workers are hindered, threatened or simply killed. All operations were suspended for a week after the assassination of a Catholic Relief Service worker in April 1995. Following the killing of an employee of the International Committee of the Red Cross (ICRC) and the wounding of several colleagues, aid efforts were halted for almost the entire north-eastern population. Finally, on 4 June 1996, three expatriate ICRC workers were brutally assassinated. Aid agencies symbolically suspended their work for a week, but many may now pull out permanently.

By the end of 1995, some humanitarian agencies were ready to break their silence. The political world followed suit with countless diplomatic initiatives by ex-presidents Julius Nyerere of Tanzania and Jimmy Carter, former Anglican archbishop Desmond Tutu of South Africa, Bernard Koushner of France and others. This growing attention culminated in UN Secretary-General Boutros Boutros Ghali's proposal for military intervention. Yet such initiatives have proved insufficient.

THE NEW IMPOTENCE

Just like Liberia and a growing number of other complex emergencies, Burundi is one of those crises for which a traditional humanitarian response has become increasingly inadequate.

Most crucially, in this intricate humanitarian context, the needs are less apparent. There are no hordes of emaciated refugees as seen in the camps of Goma here. The Burundi population in danger comprises a conglomerate of villagers who are massacred and then disregarded. Almost half a million destitute civilians are in hiding, out of reach of relief, with malnutrition silently creeping in. Several hundred thousands live in forgotten displaced settlements. Clinics and hospitals are closing one by one.

The dragging conflict has created a increasingly fragmented demographic mosaic, severely complicating aid strategies. Tutsis live either in the big Hutu-free cities or in some 180 rural displaced camps close to the military barracks. The Hutus, driven out of their towns and villages in conflict zones, seek shelter in the jungle or hide in the houses of friends. In this maze of varying needs and circumstances, aid strategies have become utterly perplexing. Uniform national strategies are impossible, while donors are unable to define a clear policy.

Aid workers in Burundi have to function in a context of high insecurity. Protection is nearly impossible because its source is unknown, unexpected and hidden. Unpunished ambushes, muggings, assaults and murder are part of their daily environment.

Gradual decay of the state and central authority have caused the disintegration of institutions that used to provide a safety net. The few public services that remain are turned over to humanitarian agencies. This allows the government to channel more means towards its war economy. In its national budget for 1996, Burundi sharply increased the defence budget at the expense of health care and education. Half of the country's hospitals are without doctors. International medical relief agencies are expected to fill the gaps in supply, logistics and personnel. If they refuse, it is to the detriment of the civilian population. Helpless Burundians are the mercy of the constant jostling between the warring parties and the aid agencies, the latter often forced to make a choice between ideological compromise and abandonment of the civilian populations. This dilemma increases as the strife drags on. Relief observers believe that the Burundi conflict may last for many more years, during which the dependency on international aid and the indirect support of the war machine will reach the obscene.

In the same manner, international humanitarian aid in Burundi, furnished by a disoriented donor community alongside ineffective policy initiatives, risks becoming entangled in the political agenda. Hence, the formation of a complex of human rights, humanitarian aid, conflict-prevention and diplomacy and national security policies.

Not surprisingly, the confusion of the humanitarian community is largely the consequence of a growing political incapacity to resolve or even directly influence complex political crises that disregard the traditional rules of the game. The Burundi conflict is amongst the most enigmatic.

The conflict's long historical foreplay has become intrinsically interwoven with today's solutions. Mediators need to take extremely complicated parameters into account in order to understand and positively influence the situation.

The gruesome Rwandan precedent remains a tangible cloud looming over each Burundian. Economic sanctions are less of a threat to the Burundi politician than the prospect of wholesale slaughter of his loved ones. Strategies of self-defence and pre-emptive strikes prevail, broadly condemned by the international community. Burundians are incited by extremist pamphlets to kill before getting killed and no public sermon will stop them. Human rights are buried under the justifications of *raisons d'état*, self-preservation and revenge.

Up till now, most key contenders have refused to recognise the problem. Even the identities of those responsible for manipulating the situation continue to remain hidden.

All the above contribute towards making attempts at mediation enormously problematic. While the international community grapples with finding an appropriate strategy, the humanitarian situation decays, undisturbed by frantic diplomatic initiatives and frustrated relief agencies. As with Liberia, Sierra Leone and Somalia, Burundi waits forlornly for the international community to make up its mind.

Lucas van den Broek, Belgian Programme
Coordinator, MSF, Burundi

MSF and other NGOs had actively participated with the Ministry of Rehabilitation and the UN in setting up Operation Return, a plan designed to return the displaced to their villages in as peaceful a manner as possible, at the end of 1994. But in early 1995, we expressed reservations about the government's proposed plan to dismantle the camps, since it did not seem to provide adequate guarantees for the safety of the displaced. Perhaps this was the 'obstruction' some journalists referred to. We could not keep quiet our well-founded suspicions that the government forces would execute Hutu leaders found in the camps. Nor could we sit quietly and not demand that transit centres and clear judicial procedures be laid out in advance of the dissolution of the camps. We could not stand by and by our silence sanction revenge killings for the genocide that had taken place. But each stand we took caused controversy and tested our position as a humanitarian organisation.

Humanitarian organisations faced similar dilemmas concerning the prison and judicial systems in Rwanda. At present, 73,000 people remain imprisoned in Rwandan jails built to hold a little over 30,000. Any improvements

Plate 10.3
Refugees line up to
collect water from a
relief agency water
tanker at the
Kibumba Refugee
Camp, Goma, Zaire.
Photograph:
Esmeralda de Vries

the Ministry of Justice might make in these prisons would have no effect on the municipal holds and police stations, which remain positively medieval and inhumane. Although it is difficult to quantify accurately, intolerable prison conditions have driven the death rate to catastrophic levels. The rate of arrest, once at 300 a week, has zoomed back up to over 1,000, making it difficult to believe due process is possible. And, considering the snail's pace at which judicial reconstruction and the establishment of a legislative structure are proceeding, it seems likely that the regime's hard-liners intend endless pre-trial imprisonment to serve as the punishment.

At the beginning of 1995, MSF published a report condemning the deplorable conditions at the Gitarama prison. The prison had a capacity of 400, yet 7,000 prisoners were being held there. Over the course of nine months, one out of every eight had died as a direct result of the conditions. The Rwandan government responded by requesting international aid to construct a special detention centre for children under 14. We were left with a dilemma. We could justifiably refuse to help, on the grounds that the project was wholly inadequate and we did not wish to aid a patently repressive government. However, our medical concerns outweighed these considerations and we chose to help evacuate the children from these ghettos

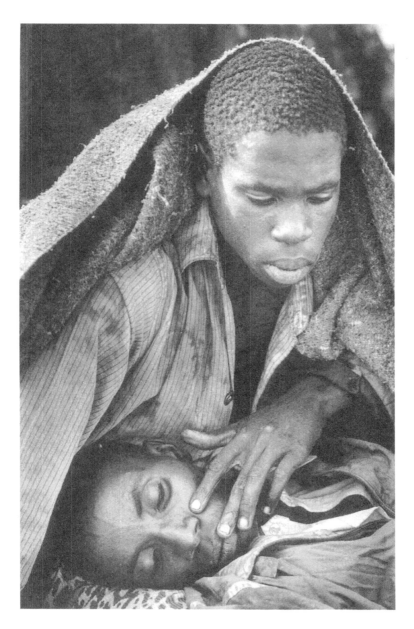

Plate 10.4
Cholera has killed hundreds, despite the existence of a relatively simple cure. Goma, Zaire.
Photograph: Remco Bohle

Plate 10.5
Refugees carry water
at Benaco Camp in
Tanzania. Photograph:
Harrie Timmermans

of death. So we were simultaneously confronted with the problem of aid
in the prisons and a commitment to the judicial process, two new fields of
intervention which pose delicate and difficult issues for an organisation such
as ours.

We were similarly challenged in the debate over the re-establishment of
a judicial system. Like others, we supported this enterprise, particularly
through the 'Citizens' Network', by supplying some of the necessary funds
and staff. But the effort seemed fruitless. The establishment of a law forbid-
ding genocide, the indispensable legal reference of a functioning system of
justice, has yet to pass in parliament. Thousands of weekly arrests have
continued, based purely on ethnic origin rather than any legal indictment.
Of course, some may point out the immensity of the task and the impos-
sibility of preserving fundamental principles of justice (under normal
conditions of due process, it would take over ten years to bring all of these
people to trial), and it is probably true that post-genocide Rwanda cannot
afford such a judicial luxury. But we must admit that, two years later,
progress in this domain is less than impressive and the facts – like the
August 1995 dismissal of several moderate government ministers – often
contradict the reassuring speeches of Rwandan authorities.

It is not easy for humanitarian organisations to criticise the Rwandan
government these days. In the eyes of the world, the FPR remains, above

all, the party which put an end to the genocide of the Tutsis and wiped out Habyarimana's murderous regime. Burdened with its own guilt, the international community is incapable of seeing them as anything less than the vanquishers of Evil. And now, how can they be anything else but the defenders of Good? To perceive the Rwandan government in perspective, to denounce their totalitarian excesses, to condemn their sometimes blood-thirsty behaviour seems – often to the world, and even sometimes to us – iconoclastic, sacrilegious. Despite everything, and even if, in the extremely polarised Rwandan context, stigmatising the atrocities of one is fast inter-preted as absolving the genocide of the other, we at MSF remain convinced that it is our duty to denounce the actions that widen the gulf between a fear-stricken population and its political representatives and that enforce the split between the Rwandan government's hard-liners and those who seek a just and peaceful solution to the crisis. We must stand by our beliefs even if there are risks involved, as when MSF's French team was expelled from Rwanda in December 1995.

Plate 10.6
Prisons such as this detention centre in Kigali, Rwanda, are severely overcrowded and lacking in facilities.
Photograph: Sebastião Salgado

FINDING NEW HUMANITARIAN PATHS

In such a context, humanitarian organisations are all too aware of their limits: limits to their interventions, since they haven't the power to prevent the APR's brutality or arrest the leaders in the camps; limits to their principles, since they cannot say that all those they assist are victims – indeed, some may even be assassins; limits to their political commitments, since the arguments they use to try to stop the killing of innocent civilians are often used by the adversary to legitimise their actions. What can a humanitarian organisation do in the middle of a war, when it has decided that it is not enough simply to try to humanise it? MSF has explored several new avenues, from promoting the justice system and protection of prisoners to monitoring what occurs in the camps. Some must be considered failures, harshly sanctioned by our expulsion or, worse still, by our voluntary withdrawal. But some have paved the way for new forms of intervention, lending substance to our vision of the future humanitarian role we would like to see evolve. If the impact of these new approaches is still, at this point, difficult to evaluate, and if they sometimes lead to thorny moral dilemmas, for us they nevertheless correspond to a firm will to make our humanitarian acts consistent with our ethical and moral credo without making the innocent, morally and physically distressed civilians hostages and involuntary victims of our policy.

The Rwandan drama, no doubt more than any other crisis, has revealed in all its naked cruelty the ultimate challenge to the humanitarian organisation at this turn of the century. The breadth and violence of our internal discussions, within our teams, in the field as well as at our home base, have proved it is a tremendous challenge to find a position of consensus. Nevertheless, we must find the courage and the strength to go on struggling to find new forms of action, new humanitarian directions. In that effort, we can at least recognise our fragility and the ambiguities we face in order to develop a mature humanitarian policy.

11

THE SUDAN
DYING A SLOW DEATH

Marelle Hart and Stephan van Praet

•

Just as it had used droughts to conceal the real humanitarian plight in the 1980s, Sudan has succeeded in diverting the attention of the outside world from the ongoing miseries of war during the 1990s. Since 1983, an estimated 1.4 million people have died in Africa's largest nation as a result of war and natural disasters. A further 3.5 million are currently displaced within Sudan's borders. As reflected by their demands for basic sustenance (if we don't receive food, we don't need your latrines), internal refugees in the Khartoum area, the majority of whom are of black African origin, depend on relief aid for basic survival.

Since its independence from Britain in 1956, Sudan has been caught up in an intermittent but continuously devastating civil war that has ravaged mainly its southern region. The Khartoum government allows humanitarian assistance, but only under certain conditions and only to alleviate some of the worst of its relief predicaments. International instruments designed to ensure the protection of the most vulnerable civilian communities are inadequate for the essential protection of those who are internally displaced. Most are barely adhered to, if at all. Struggling as it is with the ambiguity of its Afro-Arab heritage, Sudan today largely exists in a state of limbo; a vacuum in which numerous civilians enjoy few basic rights.

LINES OF DIVISION

The outbreak of civil war just before independence from Britain was not first and foremost about religion. Since then, however, the conflict has become largely a matter – at least from the points of view of the government and proselytising religious organisations – of Islam versus Christianity and traditional beliefs. Most of the press coverage on Sudan adheres to this interpretation and describes the war in Sudan as a religious war between the Arab Muslim north and the Christian and animist African south. But the north and south are not two monolithic blocks opposed to one another. They are both composed of several distinct ethnic and cultural groups. Northern populations like the Fur, Nuba and Beja, albeit mainly Muslim,

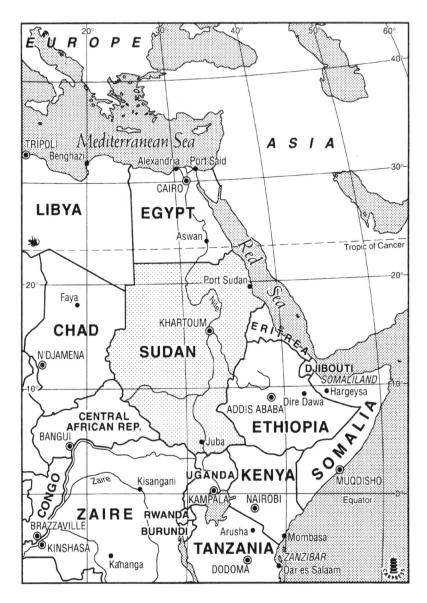

Map 11.1 East Africa: Sudan is situated south of Egypt, with Chad lying to its west and Ethiopia to its east.

are much more 'African' than the rhetoric which divides the country on the basis of religion tries to prove. Although the north remains culturally and logistically linked to the Arab world, only a minority of those living there can claim to be true Arabs. The ethnic diversity of the 27 million inhabitants of Sudan gives the country a colourful and richly varied culture.

The south is one of most remote regions of Africa. The sprawling Nile marshes known as the Sudd and the lack of transport facilities have obliged this region to seek its primary links with east and central Africa rather than Khartoum. It is not without reason that most of the southern region's trade, food supplies and humanitarian relief are brought in through Kenya and Uganda. During the colonial era, the southern and western parts of this enormous country remained economically and politically neglected.

Most of Sudan's development, mainly trade and large-scale agriculture but also rail, road and shipping links, focused on the northern and Red Sea areas. It has proved no different since independence. When it became apparent that the south commanded considerable oil and other mineral wealth, coupled with vast water resources, the north found more of a reason not to relinquish any of its control there. Limited economic links between north and south were disrupted by war and other forms of insecurity. Nomadic movements and long-distance trade became risky or impossible. Today, mobility in the south is often limited to an involuntary movement in search of food, shelter or security. Normal links between urban centres and the rural hinterland have also disappeared. Government garrison towns throughout the south have to be supplied by military convoy or by air.

The political division remains the biggest problem. After 16 years of sanguine war, the Khartoum government gave regional autonomy to the south under the 1972 Addis Ababa accords. This offered southern Sudan relative independence for just over a decade. In 1983 Nimeiry drew up new administrative divisions and introduced Shari'a law, which included non-Muslims. Cross-limb amputations: the removal of a left foot and a right hand, were introduced among other forms of physical punishments. Matters did not essentially improve following Nimeiry's overthrow in April 1985, despite efforts at renewed democracy and religious freedom.

On the eve of a peace agreement between the government of Sadiq el Mahdi and the Sudanese People's Liberation Movement (SPLM), the military-backed National Islamic Front took power in June 1989. Repression, not only in the south but also against opposition groups in the north, grew worse. During the early 1990s, however, the southern opposition movement found itself weakened by the fall of the Mengistu regime in Ethiopia, and by its own internal political and ethnic rivalries. The SPLA (Sudan People's Liberation Army) split into Mainstream and United elements, now known as the SPLM and the Southern Sudanese Independence Movement – the SSIM. While the Khartoum regime of the north claims to be involved in a jihad or holy war, the armed opposition in the north and south says it is fighting a political struggle for a multi-ethnic and democratic Sudan. The result is tragic: an endless civil war whose devastation affects the institutions of the entire Sudanese nation and the lives of ordinary citizens in both the north and the south.

Plate 11.1
Women in Mwele camp for the internally displaced refugees outside Khartoum, September 1991.
Photograph: Marie José Sondeijker

HUMAN RIGHTS ABUSES: A STATE WITHOUT PROTECTION

'Because human rights protection is not a central concern or function of most other relief and development agencies, and because human rights bodies are not yet fully operational and often not present, protection for the internally displaced is one of the most pressing gaps in the international system,' maintains Francis Deng, a Sudanese national and Special Representative of the Secretary-General for internally displaced persons in 1995.

The most painful and tragic events occur when it is the government itself which denies its citizens protection or, worse, is directly implicated in major violations of human rights despite its legal and moral obligation to protect them. Article 14 of the constitution of 8 May 1973 declares that Sudanese society is based on the principles of liberty, equality and justice. Unfortunately for the citizens of Sudan, that constitution was abolished in 1989 by the present regime of General Omar Al Bashir. Today, if all the reports on human rights violations in Sudan were laid side by side, they would be long enough to form a bridge across the Nile.

Sudan has signed international conventions that oblige the government to protect its citizens, such as the Slavery Convention (1926), the Universal Declaration of Human Rights (1948), the International Convention on the Elimination of Racial Discrimination (1965), the Universal Islamic Declaration and the African Charter on Human and Peoples' Rights (1981). However, the successive governments of Nimeiry, Dahad, Sadiq el Madhi and especially Bashir have had at least one feature in common: a lack of protection for their citizens. In contrast to the warm hospitality offered by the Sudanese people towards foreigners, respect for the rights of their own nationals seems to have evaporated in the desert wind.

According to Amnesty International, Human Rights Watch, African Rights and other human rights organisations, government soldiers and militia frequently round up children and send them to secret camps around Khartoum as well to the eastern and western parts of the country for religious and military training. Alternatively they are sent, literally as slaves, to serve as domestic servants with Arab families in both the Sudan and the Middle East. Humanitarian relief agencies have been told of slavery cases by civilians with whom they work in Sudan, and sources as far away as Kuwait and Riyadh have reported forms of slavery involving Sudanese citizens.

Slavery is very much a reality in Sudan today. Dispossessed farmers and villagers, both men and women, often become forced labourers in government agricultural schemes or on private farms owned by regime officials. Many others, deprived of their grazing lands for their livestock, find themselves trapped in displacement camps in abject poverty, robbed of their dignity and left with few or no means for survival. The Khartoum authorities deny all these reports. They claim the children are provided with food and education and that there is no state-sponsored kidnapping. The use and abuse of civilians as pawns, however, is not restricted to government forces. Young men and women are occasionally conscripted by all the armed parties. Both government and rebel forces press-gang people for short military training programmes. They are then sent off to war as members of local militia. Nearly one million civilians reportedly serve in Khartoum's Popular Defence Force (PDF) in northern Sudan, a government militia that ostensibly claims to embrace a glorious and heavenly path to martyrdom as it marches into battle under the banners of Islam.

ISOLATED POPULATIONS

The central provinces (South Kordofan, Upper Nile and Bahr el Ghazal) have become isolated as a consequence of the general state of insecurity and lack of infrastructure. Huge tracts of land, such as the rural areas inhabited by the Nuba, a cluster of ethnic groups in the central mountains of Sudan, have been completely cut off from the outside world for nearly a decade. The populations of the 'transitional zone' between north and south Sudan have long practised a strong tradition of religious tolerance. Muslims, Christians and the holders of traditional beliefs were often members of the same family. However, because of the manipulation of religion for political ends, coupled with racial discrimination, this tolerance has been threatened by desperate confusion and bitter resentment. The government blames village elders whenever the youths of an area join the SPLA. Government troops then wreak havoc by burning churches as well as mosques. Civilian targets, too, are indiscriminately bombed by the governmental air force. Many rebels or uncontrolled armed groups have gone on the rampage by killing and looting and, like the government, do not seem to differentiate between religious groups.

Because these huge enclaves are largely cut off from regular contact with other parts of the country, reports of these indiscriminate attacks are rarely heard. In the Nuba mountains, the government's military campaign, involving the relentless destruction of villages and the forced relocation of populations, is described by human rights groups such African Rights as genocide.

INTERNALLY DISPLACED

Towards the end of the 1980s, large-scale migrations of mainly black African refugees began to overwhelm Khartoum. Those from the south fled the war, and civilians from the western regions fled the effects of the 1984–5 famine there. While those from the western part of the country and others seeking economic improvement were called squatters, southerners were referred to

as 'displaced'. Treated little better than dogs, if not worse, as one European aid official put it, they completely lacked any resettlement rights.

During the 1986–9 administration of Prime Minister Sadiq el Mahdi, a Sudanese Arab who, despite his Oxford education, favoured Shari'a and had little time to consider the rights of his southern African brethren, thousands of internal refugees were simply loaded on to trucks and dumped outside Khartoum. Such heavy-handed treatment continues under the current NIF, which has adopted a more determined, and organised, form of discrimination known as 'town planning'.

In 1992 the NIF government announced its intention to prepare villages for returnees who were either liberated or who had left SPLA areas on a voluntary basis. The authorities referred to such locations as 'peace villages'

Map 11.2 Sudan: Médecins Sans Frontières project sites are located at Leer (north of Rumbeck), Duar (north-east of Waw), Khartoum, Al Qadarif, Dilling (south of An Nahud), Akobo (north-east of Bor), Waw and Yambiyo.

Plate 11.2
Women in North Darfur Province. Photograph: Marie José Sondeijker

just as local militia and police are dubbed 'popular forces', while the process of eliminating opponents was known as developing 'peace from within'. The reality, however, was far more insidious. The 'returnee' villagers were part of a large-scale forced relocation programme, a plan for social engineering in the form of political repression. Men, women and children were screened and separated, often held in conditions more reminiscent of prison camps than relocation centres. Many of these 'villages' have become little more than forced labour camps to provide free or cheap workers to mechanised farms. The so-called 'ghost houses' (a grim parody of the term 'guest house') function as secret detention centres run by different security organisations for torturing or otherwise intimidating suspected political opponents. 'The main fear is the spread of security networks,' said one Sudanese doctor formerly with the Ministry of Health in Khartoum. 'Most children are indoors, but if your son disappears you wouldn't know where to look for him.' The number of families that have been forced to seek refuge in the camps, with other families, or abroad, is beyond calculation. Also incalculable is the number of families torn apart both by war and slavery.

Ostensibly for industrial, agricultural and road purposes, the government has taken brutal measures to clear squatter camps. Armed police and soldiers move into areas using teargas and demolish homes with bulldozers. The victims are neither compensated nor provided with official sites to which they can move. Relocated people are simply left on open ground, often on the extreme edges of Khartoum, with no electricity, sanitation facilities, or even water. They are forced to rebuild their houses from scratch. Removals often take place in winter, forcing parents to dig holes in the hard ground so that their children can get some meagre shelter from the biting wind. 'The main reason is to keep the displaced population destabilised,' says Ellen Colthoff, a Dutch field director of Médecins Sans Frontières (MSF).

Among the country's 3.5 million internal refugees an estimated 1.9 million have been forced to encamp in and around the capital. The four official camps for the displaced are located from 25 to 40 kilometres outside downtown Khartoum. These include Jebel Aulia and Mayo Farms to the south, and Al Salaam and Wad el Bashir to the west. Considered by most Sudanese migrants or internal refugees as places of last resort, only a small percentage of the total number of the displaced find themselves there. As the desert winds whip round and through their flimsy *tukul* shelters, there is little reason to go out. Thousands go without water because the government will not repair the few existing water facilities. Many camp inhabitants remain in a permanent state of hunger, lacking adequate food stocks of their own and unable to rely on the irregular relief agency supplies. 'People are just lying in their houses, with no food and no money,' says a Sudanese hygiene educator employed by MSF to monitor family conditions.

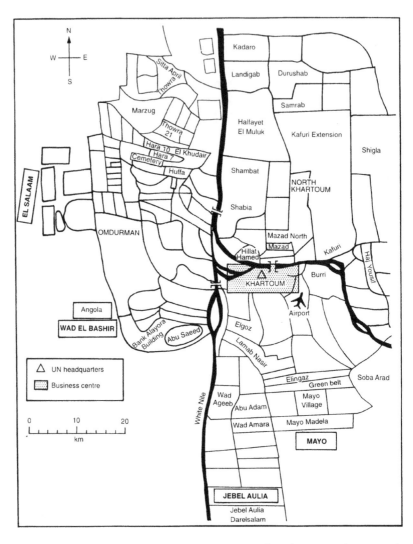

Map 11.3
Khartoum: Sudan's capital city has become a focus for the country's internally displaced. Fleeing war and famine, they have established camps on the outskirts of the city. There are four official displaced camps at Jebel Aulia, Mayo Farms, Al Salaam and Wad el Bashir, where their inhabitants live in dire conditions.

In October 1995, the government demolished without warning an estimated 450 houses and 176 MSF-constructed latrines in Al Salaam camp, ostensibly because of boundary realignments. The victims were mainly Nubas. Behind the levelled houses, the camp looked out on to the horizon. Apparently, there was no space in an empty desert for these people who were left with nothing. At the time of writing, an estimated 1,000 families from the Fithab sector of Khartoum and another 10,000 families from the Angola squatting quarter have been rebuilding small structures on these wide sandy plains. Their houses were all destroyed between October 1995 and January 1996. Pushed around, with nowhere to go, many of those displaced remain squatting.

The government fears, or at least claims to fear, that the southern movements will one day use the displaced to organise an attack on Khartoum. 'All it takes is a quick trip to the camps to see that they are quite harmless,' says Robert Painter, chief of the United Nations' emergency unit in Khartoum. As for the refugees, they fear that the government is pushing them further and further into the desert in order, eventually, to exterminate them. 'Nature will take care of that. They are not giving them anything,' noted Maja, a Sudanese doctor.

The Khartoum authorities are seeking to isolate these unwanted displaced and discourage them from building a new life by driving them well away from the capital. As displaced persons, or refugees within the territorial confines of their own country, it is their own government and not the United Nations High Commissioner for Refugees (UNHCR) that bears the primary responsibility for their well-being. Any international organisation that seeks to defend the physical safety and fundamental rights of these internal refugees is barred from operating in Sudan. Local humanitarian groups can do nothing because of government pressures, and security forces have simply ignored protests by concerned citizens.

INTERNATIONAL RESPONSE

The rights of the Sudanese people to survive, to feed themselves, to live together, to practise their culture and their religion, to assist each other, to gather together in groups, to choose their own way of life – these are rights that have no weight in the forum of international politics and the strategies of international companies. Although concern may occasionally have been expressed, it was never backed up by any show of real intent to influence the situation. Indeed, a second bridge could be built, over the same stretch of Nile, from the number of resolutions condemning human rights violations in Sudan issued by individual countries or international organisations, such as the United States Congress, the European Parliament, the Commission of the European Union and the UN. There would then be two bridges going nowhere.

It was, for example, a praiseworthy statement that the European Union made on 21 February 1994, inviting the Sudanese government to look for negotiated solutions and to encourage diplomatic efforts – praiseworthy, but totally ineffective.

On 16 March 1994, the EU declared an embargo on the export of weapons, munitions and military equipment to Sudan, but without any effective controls to ensure it was respected. Although the European Parliament adopted a resolution in December 1994 that condemned human rights violations by government forces and rebel factions, and called for the reinforcement of sanctions against the Khartoum regime, some member

Never forget

We were working in southern Sudan, a part of the world which was under siege, plagued by famine, and away from which the government was trying to drive out the Nuer and the Dinka with land and air attacks.

Aid organisations were having trouble bringing in food, as they could not get permission to fly in supplies. At the time, the government denied (and even now continues to deny) the existence of a deadly epidemic of Kala Azar. Without treatment, 95 per cent of the population would have died. The word 'genocide' springs to mind . . .

We continued to work in isolation, using a faltering supply line. We worked ourselves into the ground, fighting to make our programme succeed. Cut off as we were, we suffered emotionally from the enormous pressures of work, the residual 15 per cent mortality rate and the inability to do more for want of human and material resources. People died around us in great numbers, even in our compound, as we could only watch.

There came a time when we could stand it no longer. Exhausted, we ran around in a daze, waiting for a plane to pick us up.

The people of southern Sudan were fantastic. We were outsiders who were giving up and returning to the safety of our country and our families and friends who would nurse us back to emotional health, however arduous that proved to be. But the people we had to leave behind had no escape, and no choice but to go on.

I could not understand how they could take it, and I told them how privileged I felt to have a safe home to go to and to be able to get away from the sorrow of Sudan. I told them how sorry I was to have to leave them, and how guilty I felt.

Bad though their situation was, they calmed me down, explaining that they accepted their lot and saw it as something they wanted to, needed to and were obliged to get through.

They asked only one thing of us: that we go home, rest and recuperate, but that we talk about them and their situation to our families, friends and compatriots – that we tell the world about them so that it didn't forget them.

This is why I will never stop talking about them.

Mieke Knoppers, Dutch medical doctor, MSF, Southern Sudan

states, such as France, still continue to back the commercial activities of private companies and semi-official institutions.

In 1995, an agreement was signed between Operation Lifeline Sudan (OLS, a tripartite agreement set up in 1989 between the UN, the government of Sudan and the resistance movements in the south), the SPLM and SSIM, laying down new ground rules, including respect for the principles of several international conventions (Convention of the Rights of the Child,

1989; Geneva Conventions of 1949 and the Additional Protocols of 1977). Although it is a positive sign that OLS has now integrated the promotion of humanitarian principles into its programmes, why did it take so long to get around to this?

It is to be hoped that these principles will be put into practice and have a real impact on all sides at war in the frontline villages. But hope alone will not be enough to tackle human rights violations and the issue of impunity. It is essential that there is human rights monitoring in the field in order to have at least an idea of what is really happening. But such monitors are officially refused entry to Sudan and even the Special Rapporteur of the UN Human Rights Commission, Gaspar Biro, was considered *persona non grata* for years by the government.

Occasional intermediaries, such as former US President Jimmy Carter, who mediated a temporary cease-fire during the rainy season in 1995, have tried without lasting success to open new diplomatic channels. A series of peace talks jointly organised by the four countries of the region – Eritrea, Ethiopia, Kenya and Uganda – have collapsed because of mutual suspicion. It took the assassination attempt on a foreign president, Hosni Mubarak of Egypt, to provoke the international community into threatening Sudan with economic sanctions – without results. The lack of international attention and Sudan's consequent international isolation has allowed large-scale violations to occur without any substantive international response.

THE CHANGING HUMANITARIAN RESPONSE IN SOUTHERN SUDAN

It is estimated that 1.4 million people have died in Sudan since 1983 as a result of war and natural disasters, and around 4.5 million are currently affected today. The Khartoum government has allowed humanitarian intervention – but only under certain conditions – to alleviate some of the worst problems, and UN agencies, the ICRC (International Committee of the Red Cross), and international and local NGOs (non-government organisations) have achieved some impressive results.

Relief assistance is mainly provided under the umbrella of OLS. This OLS agreement allows the humanitarian agencies to operate in relative security and to carry out cross-border operations. OLS has made it possible to bring in aid by air to locations in the south that would not otherwise be accessible. As the war drags on, agencies bring food to starving populations via an air-bridge operating either from Khartoum or from the Kenyan village of Lokichokio, the logistics base for the southern sector. This former Tukana village near the border with Sudan has been transformed over the past ten years into a huge humanitarian base, the waiting-room for operations in southern Sudan. The airstrip has been given a tarmac surface and a control tower, turning it into Kenya's third most important airport.

Around 40 international relief agencies implement programmes in agreement with the local authorities in the north or with the SRRA (Sudan Relief and Rehabilitation Association) or RASS (Relief Association for Southern Sudan), the humanitarian branches of the two main armed movements in the south.

In areas where a degree of stability prevails, it has been possible to implement large-scale programmes to tackle major endemic tropical diseases such as malaria, Kala Azar and tuberculosis, and recurrent epidemic diseases like measles. A temporary cease-fire, brokered by the Carter Center during the rainy season in 1995, made it possible to step up efforts to eliminate guinea-worm infection in the region and also onchocerciasis. However, the four months it allowed were clearly not long enough and several areas were not covered by the cease-fire. Currently, a cholera epidemic has already killed over 700 people in southern Sudan and there is a great risk that it will spread further north into the Nile valley. The question of accessibility for relief agencies is more than a symbolic one, it is a priority in the struggle for life when villages are losing half of their populations to a curable disease. The present negotiations for unconditional access to prevent a further spread of deadly epidemics represents the real battlefield between politics and humanitarianism. The failure of these negotiations would mean a deliberate strategy of non-assistance for people in need.

Over the years, OLS has slowly adapted the type of relief it offers to correspond to the chronic nature of the conflict. Thus the focus has switched from providing food towards the provision of primary education, water and sanitation facilities, and household food security. A multi-sectorial approach now promotes primary health care, emergency preparedness and the development of Sudanese relief organisations. Relief programmes are also paying increasing attention to training Sudanese health workers, teachers and veterinarians.

The ecological vulnerability of the region and the chronically disturbed state of the local economy as a result of years of conflict makes the population extremely vulnerable to droughts or floods. Although the reserves traditionally held in case of such natural disturbances have long since been used up, farming and livestock activities remain essential for the survival of the people.

Despite the civil war, southern Sudan is not entirely given over to terror and fighting: the region represents a patchwork of areas of relative calm and others which remain more tense. There are large areas at a distance from the main garrison towns that remain quite peaceful and where the people are self-sufficient enough to survive.

Relief programmes in these areas therefore rightly integrate activities intended to have a long-term impact with those which try to improve on existing resources and abilities in order to deal with the crisis situation. A striking example is the set of medical guidelines drawn up in cooperation

Plate 11.3
A couple waits by a tree in North Darfur Province. Photograph: Marie José Sondeijker

Plate 11.4
Patient suffering from
Kala Azar disease,
southern Sudan.
Photograph: Remco
Bohle

with health staff from the different southern factions. Unfortunately, it is extremely difficult to find donors to fund programmes aiming at this kind of rehabilitation and development in conflict situations. In Sudan, this means that many valuable programmes aiming to reinforce the local infrastructure are not given priority status and many initiatives that target beyond immediate needs are unable to find backers.

LIMITATIONS ON RELIEF WORK

Hundreds of thousands of people in Upper Nile, Bahr el Ghazal and southern Kordofan, isolated by the general insecurity and lack of transport, are deprived of the most basic commodities. Salt, tea, sugar, soap, clothes, shoes, pens and even paper became so scarce that many of the Korans and Bibles sent in by religious organisations literally went up in smoke as they were used to roll the locally grown tobacco into cigarettes. Officially, relief cannot be sent into the Nuba mountains in southern Kordofan as the government does not consider this region of northern Sudan to be a war zone and has refused OLS permission to operate there.

There are relief programmes outside the OLS framework, although their scope is limited. The ICRC has its own agreements with the government and the warring parties, and several NGOs – most of which are no longer

allowed to work officially in Sudan – started parallel relief activities under agreements worked out directly with individual churches and factions.

However, these activities are mostly limited to the regions bordering Kenya and Uganda, since the remoteness of the country and the logistical difficulties in regard to overland transport make relief operations extremely expensive. Many NGOs therefore cannot afford to work independently of OLS, which provides free air transport for personnel and cargo. However, if they benefit from the advantages of using OLS, they also have to accept OLS regulations and restrictions or else they risk being unable to operate.

The conditions imposed on humanitarian aid reached unbelievable levels for those agencies working in northern Sudan, necessitating administrative procedures that have given Khartoum a reputation as a capital city of obstructive red tape. A variety of official measures continuously challenge the basic principles of many international agencies: pressure to work through local NGOs – often selected or imposed by the government – difficulties in obtaining travel permits, insistence that specific personnel are taken on, etc. In order to ensure access to the populations in need, the principles of impartiality, independence and accountability are continuously stretched to the limit.

As aid professionals are only too aware, the humanitarian organisations cannot be expected to go on providing relief indefinitely. Aid fatigue makes

Plate 11.5
Refugees stand in line for food rations, southern Sudan.
Photograph: Chris Steele Perkins

it increasingly difficult to generate sustained commitments of resources for emergency assistance. While some relief agencies have been expelled from Sudan, others have left because of incontestable interference with their aid programmes.

At times, relations between Sudanese nationals and expatriate aid workers are strained, the former suffering from wounded pride, the latter unwitting arrogance. Relief workers often place a higher premium on getting the emergency job done than collaborating with their Sudanese counterparts. Caught up in their work, those providing assistance often find themselves involved in brokering peace, even mediating directly between the warring parties. Short-term humanitarian agendas interconnect with longer-term peace objectives. The discussion of one nearly always involves the other. Humanitarian agencies, however, are neither mandated nor equipped to carry out monitoring or conflict resolution activities. Lingering needs in the Sudan call for a sweeping remedy to address the underlying causes of the conflict.

OPERATION LIFELINE SUDAN UNDER PRESSURE

A major review of OLS is taking place in 1996. There are a lot of questions to be raised regarding operational efficiency, cost efficiency, bureaucracy, co-ordination, the effects that OLS has had on the dynamics of the conflict itself, the limits of neutrality, and other topics. Everybody involved in humanitarian work in Sudan is interested in what will result from this review.

One of the major issues is the basic principle of working under an agreement signed by a number of different warring parties. Although the OLS agreement is a unique instrument that has saved the lives of thousands, the indirect consequences that have followed on from it have had great significance for humanitarian work. By agreeing to accept conditions laid down by the belligerents, the humanitarian agencies have made themselves subject to the will of those who could manipulate aid to their own ends.

Thus by accepting, for example, regular flight bans, restrictions in regard to the use of certain types of aeroplanes, the lack of access to areas declared unaffected by the war, such as the Nuba mountains, or to regions suddenly affected by major epidemics, it is clear that OLS depends on unilateral government decisions to be able to carry out its work. To avoid a total collapse of the agreement, OLS is prepared to sacrifice some endangered communities and make concessions on basic principles of humanitarianism, for example the freedom of access to all populations in danger. In addition, its own internal security procedures have resulted in OLS regularly recalling NGOs from the field, obliging them to abandon starving people. Thus staff are left waiting in frustration in Khartoum or Lokichokio to receive the

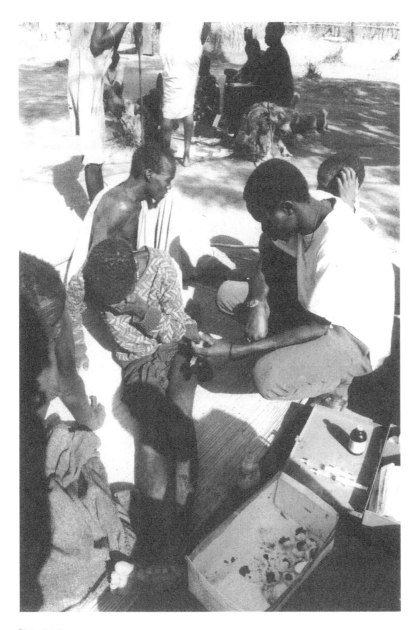

Plate 11.6
A Kala Azar patient receives treatment, southern Sudan. Photograph: Peter Slavenburg

Evidence of non-protection of civilians in Sudan:
some examples

March 1983: Mutiny by the 105th Battalion of the Sudanese Army unleashes the country's second civil war. During the first years of the war, an important famine strikes northern and western Sudan, prompting a major international relief effort. There are systematic raids against and large-scale killings of Dinkas in northern Bahr el Ghazal by armed militias.

March 1987: More than 1,000 Dinka civilians massacred by Rizeigat militia in railway station at El Daein.

August/September 1987: Ethnic clashes in the city of Wau result in the deaths of more than 2,000 civilians. Large-scale famine in Bahr el Ghazal and the transitional zone between north and south leads to unacceptably high mortality rates: an estimated 200,000 people die. Massive exodus to refugee camps in Ethiopia.

June 1989: Military coup in Khartoum. The parliamentary constitution, political parties and trade unions are abolished.

November 1989: Popular Defence Act legalises the para-military militia Popular Defence Force (PDF).

December 1989: MSF-chartered plane shot down shortly after take-off by a rocket fired from inside the government-held town of Aweil. Four people killed.

May 1991: The fall of the Mengistu regime in Ethiopa leads to the sudden and chaotic return of 300,000 refugees, Some groups of returnees are bombed by the Sudanese airforce.

November 1991: Start of a government campaign in the Nuba mountains leads to the burning of numerous villages and forced population relocations to so-called 'peace villages'. Systematic destruction of villages. Relocation of more than 40,000 people to northern Kordofan. Reports of forced labour in production schemes or as domestic servants.

Summer 1992: Extra-judicial killings of 300 people by government forces in the besieged town of Juba.

September 1992: A journalist and three UN staff killed in an ambush by an SPLA faction in Ame, Eastern Equatoria.

February 1993: Pope John Paul II warns Al Bashir during his visit to Khartoum that there will be no peace in Sudan without respect for human rights.

February 1994: Mundri and camps for the displaced in Nimule region were bombed. Over 100,000 local and displaced persons flee again in search of safety. Growing influx of refugees in Uganda.

July 1994: Inter-tribal fighting between SPLA factions in northern Bahr el Ghazal, leading to the destruction and looting of humanitarian relief supplies and equipment. An estimated 1,000 civilians are killed and numerous villages burned.

October 1994: One hundred civilians reportedly killed at Akot by the SSIM forces of Riak Machar and more than 5,000 families displaced.

February 1995: 250 civilians are abducted by soldiers from the village of Toror, Nuba mountains, southern Kordofan, to so-called 'peace villages' where the rural populations are concentrated for their own protection.

July 1995: A village near Ganyiel is attacked by an SPLA faction, resulting in the deaths of 210 people (including 53 women and 127 children) and the destruction of 2,000 homes.

October 1995: Report of UN Special Rapporteur of the Human Rights Commission, Gaspar Biro. Among the violations committed by the government of Sudan, he lists: extra-judicial killings, summary executions; enforced or involuntary disappearances; torture and other cruel, inhuman or degrading treatment; arbitrary arrest and detention; elements of penal legislation inconsistent with international norms; slavery, servitude, slave trade, forced labour and similar institutions and practices; violations of freedom of expression, association and peaceful assembly; violations of the rights of the child and of the rights of women; violations of freedom of movement and residence, including the right to leave or return to the country. On the rebel side, for 1995 alone, Mr Biro reported more than 40 cases of hostage-taking among relief workers, as well as systematic cattle raiding and looting; the terrorising of civilian populations (Major Kerubino Kwanyin Bol was specifically named); extra-judicial killings of presumed collaborators.

End of 1995: More restrictions are imposed by the Khartoum government on relief flights.

December 1995: 7,000 homes on the site of the Angola displaced persons settlement, west Khartoum, were suddenly and unexpectedly destroyed within a three-week period.

February 1996: The annual dry season train brought supplies to the government garrison town of Wau in Bahr el Ghazal. The train was protected by an escort of several thousand PDF troops on horseback and armed with automatic weapons. Villages up to 10 kilometres on either side of the railway track were at risk of looting and burning. The population fled in advance of their arrival. This railway was once repaired by international donor funds in order to bring humanitarian relief to the region.

Genesis of rebel movements in southern Sudan

The SPLM/A – Sudanese People's Liberation Movement/Army – claims to seek a unified, secular democratic Sudan and the right of self-determination for the populations in the south (John Garang).

August 1991: Secession of the SPLA-Nasir faction, later renamed SSIM – South Sudanese Independence Movement. Seeks separation of the south from the north (Riak Machar).

October 1994: Secession from SSIM of commander Lam Akol to create SPLA-United, which is active in northern Upper Nile province. SSIM commander Kerubino controls Gogrial area in northern Bahr el Ghazal and collaborates with government of Khartoum. Several SSIM commanders create SSIM-2 (John Luk) and make an alliance with SPLM/A. SSIM-2 commander William Nyuon killed in January 1996 by troops of Riak Machar.

February 1996: Official peace agreement between Riak Machar, Commander Kerubino and the Khartoum government.

NDA – National Democratic Alliance, a coalition of SPLM/A and several northern political parties (Democratic Union Party, Umma, Beja congress, etc.) meets regularly in Asmara, Eritrea. One of their armed branches is the SAF (Sudanese Allied Forces).

End of June 1996: Akobo recaptured by forces of Riak Machar.

---

green light in order to carry out their programmes, which are dependent on the whims of those who are directing the war game.

Humanitarian agencies working under OLS have become dependent on the capricious warring parties. Many relief workers have been attacked or taken hostage, and some of them have been killed. If they are to abide by OLS regulations, humanitarian agencies must obey and withdraw when told to do so, hoping that their programmes can manage without being re-supplied. There is no way of punishing those who break the ground rules for there are no procedures in the OLS agreement to deal with violations of human rights and international humanitarian law.

It is widely known that access to southern Sudan was obtained at the cost of diplomatic concessions and the provision of large quantities of food supplies to the government garrison towns in the south. It is also clear that the numbers of displaced persons that are used for calculating food aid requirements are, on all sides, far higher than the reality. While it must be accepted that hungry soldiers and hungry rebels are a major threat to local communities, where do you draw the line between feeding those who are in desperate need and feeding the war effort? Population displacement is often forced as part of a strategic plan: those in charge of decision-making know that the food aid will follow the people.

PREDICTIONS FOR THE FUTURE

Whatever the long-term future holds for Sudan, whether or not the country will unite or divide, the rights of minority groups and of displaced populations in northern Sudan will remain one of the key issues in Sudanese politics for today and tomorrow. It is therefore essential that questions of access, assistance and protection for these highly vulnerable communities become a priority for the humanitarian organisations. The interests of populations in danger must be the criterion for any kind of humanitarian intervention. Relief agencies are accountable to those they work for, as well as to those who provide the funding. If humanitarian aid is not linked with minimum respect for human rights and international humanitarian law, the relief aid itself is brought into question.

Thus, the problem of how to link assistance and protection without abandoning any of those in need is one of the major challenges for humanitarian relief. There must be no restrictions placed on the unconditional right of free access to all populations in need of assistance or protection, and this right must be insisted on by all humanitarian organisations at all times. Mechanisms must be created to ensure that relief workers in Sudan and elsewhere are no longer condemned to be powerless witnesses of chronic human suffering and oppression.

The international community must open its eyes and look again at the situation in Sudan. The tripartite OLS agreement may well have been the excuse for the UN to resist involving itself in the field of protection and conflict resolution, but it must now be recognised that this perspective is too narrow. It is high time that the internal situation in Sudan is taken seriously. The evidence of abuse is sufficient for an active policy to be developed that will seriously resist the efforts of the Khartoum government to strangle the entire nation. This conflict has been going on for 30 years. If we continue to block out the cries of pain, the time is not so far off when no one will be left to call out.

This chapter was translated by Mrs Alison Marschner.

MÉDECINS SANS FRONTIÈRES
WORLD-WIDE

OPERATIONAL CENTRES (*) AND DELEGATE OFFICES:

Australia
Médecins Sans Frontières
215 Abercrombie Street
Chippendale 2008 NSW
GPO Box 5141
Sydney 2001
Australia
Tel: +61 (2) 9319 3500
Fax: +61 (2) 9319 2383
E-mail: 100243.3671@compuserve.com

Austria
Médecins Sans Frontières
Arzte ohne Grenzen
Gumpendorferstrasse 95
A–1060 Wien
Austria
Tel: +43 (1) 59 60 39 00
Fax: +43 (1) 59 60 390 10
E-mail: MSF-Wien@Brussels.msf.org

Belgium*
Médecins Sans Frontières
Artsen zonder Grenzen
Rue Dupré 94
B–1090 Bruxelles
Belgium
Tel: +32 (2) 474 74 74
Fax: +32 (2) 474 75 75
E-mail: Zoom@Brussels.msf.org

Canada

Médecins Sans Frontières
Doctors Without Borders
355 Adelaide Street W, 5B
Toronto
Ontario M5V 1S2
Canada
Tel: +1 (416) 586 9820
Fax: +1 (416) 586 9821
E-mail: msfcan@passport.ca

Denmark

Médecins Sans Frontières
Laeger uden Graenser
Strandvejen 171
DK–2900 Hellerup
Denmark
Tel: +45 (39) 62 63 01
Fax: +45 (39) 40 14 92
E-mail: MSF-Copenhagen@Brussels.msf.org

France*

Médecins Sans Frontières
8, rue Saint-Sabin
F–75011 Paris
France
Tel: +33 (1) 40 21 29 29
Fax: +33 (1) 48 06 68 68
E-mail: office@Paris.msf.org

Germany

Médecins Sans Frontières
Arzte ohne Grenzen
Adenauer Allee 50
D–53113 Bonn
Germany
Tel: +49 (228) 91 46 70
Fax: +49 (228) 91 46 711
E-mail: do@Bonn.msf.org

Greece

Médecins Sans Frontières
Giatri Horis Synora
57 Stournari Street

GR–104 32 Athens
Greece
Tel: +30 (1) 52 00 500
Fax: +30 (1) 52 00 503
E-mail: Sotiris_Papasyropoulos@Athens.msf.org

The Netherlands*
Médecins Sans Frontières
Artsen zonder Grenzen
Max Euweplein 40
PO Box 10014
NL–1001 EA Amsterdam
The Netherlands
Tel: +31 (20) 52 08 700
Fax: +31 (20) 620 51 70/72
E-mail: hq@amsterdam.msf.org

Hong Kong
Médecins Sans Frontières
GPO Box 5803
N.T. Hong Kong
Tel: +852 2 338 82 77
Fax: +852 2 304 60.81
E-mail: MSFB-Hong-Kong@Brussels.msf.org

Italy
Médecins Sans Frontières
Medici senza Frontiere
Via Ostiense 6/E
I–00154 Roma
Italy
Tel: +39 (6) 57 300 900/901
Fax: +39 (6) 57 300 902
E-mail: MSF-Roma@Brussels.msf.org

Japan
Médecins Sans Frontières
Takadanobaba 3–8627
Shinjuku-Ku
Tokyo 169
Japan
Tel: +81 (3) 3366 8571/72
Fax: +81 (3) 3366 8573

Luxemburg*

Médecins Sans Frontières
70, route de Luxembourg
L–7240 Bereldange
Luxembourg
Tel: +352 33 25 15
Fax: +352 33 51 33
E-mail: office-lux@Luxembourg.msf.org

Norway

Médecins Sans Frontières
Eugenisgate
22, App 401
N–0168 Oslo
Norway
Tel: +47 (2) 259 95 90
Fax: +47 (2) 259 95 90
E-mail: MSF-Oslo@Brussels.msf.org

Spain*

Médecins Sans Frontières
Medicos sin Fronteras
Nou de la Rambla 26
E–08001 Barcelona
Spain
Tel: +34 (3) 304 61 00
Fax: +34 (3) 304 61 02
E-mail: medsf@pangea.org

Sweden

Médecins Sans Frontières
Laekare utan Graenser
Atlasgatan 14
S–11320 Stockholm
Sweden
Tel: +46 (8) 31 02 17
Fax: +46 (8) 31 42 90
E-mail: MSF-Stockholm@Brussels.msf.org

Switzerland*

Médecins Sans Frontières
12, rue du Lac
Case Postale 6090
CH–1211 Geneva 6

Switzerland
Tel: +41 (22) 849 84 84
Fax: +41 (22) 849 84 88
E-mail: office-gva@Geneva.msf.org

UK
Médecins Sans Frontières
124–132 Clerkenwell Road
London EC1R 5DL
United Kingdom
Tel: +44 (171) 713 56 00
Fax: +44 (171) 713 50 04
E-mail: office@london.msf.org

USA
Médecins Sans Frontières
Doctors Without Borders
11 East 26th Street
Suite 1904
New York, NY 10010
USA
Tel: +1 (212) 679 6800
Fax: +1 (212) 679 7016
E-mail: Dwb@newyork.msf.org

INTERNATIONAL OFFICES

Médecins Sans Frontières
Rue de la Tourelle 39
B–1040 Brussels
Belgium
Tel: +32 (2) 280 18 81
Fax: +32 (2) 280 01 73
E-mail: msf-international@bi.msf.org

UN LIAISON OFFICE

Médecins Sans Frontières
12, rue du Lac
Case Postale 6090
CH–1211 Geneva 6
Switzerland
Tel: +41 (22) 849 84 00
Fax: +41 (22) 849 84 04
E-mail: MSF-International-GVA@Geneva.msf.org

INDEX

•